Statistics for Industry, Technology, and Engineering

More information about this series at http://www.springer.com/series/15677

Shelemyahu Zacks

The Career of a Research Statistician

From Consulting to Theoretical Development

 Birkhäuser

Shelemyahu Zacks
McLean, VA, USA

ISSN 2662-5555 ISSN 2662-5563 (electronic)
Statistics for Industry, Technology, and Engineering
ISBN 978-3-030-39436-3 ISBN 978-3-030-39434-9 (eBook)
https://doi.org/10.1007/978-3-030-39434-9

Mathematics Subject Classification: 46N30, 60-02, 62-03, 90-02, 62E17, 62F10, 62P30, 62P35, 65-04, 65T40, 90B05, 90B25, 90C47

This book is published under the imprint Birkhäuser, www.birkhauser-science.com by the registered company Springer Nature Switzerland AG.
The registered company address is: Gewerbestrasse 11, 6330 Cham, Switzerland

*To my grandchildren Gabrielle, Daniel,
and Jacob Zacks*

Preface

The objective of this book is to demonstrate the strong connection between the development of statistical theory and data analysis. I have been a research statistician for over 60 years and have written numerous research papers and books. Almost all my research contributions were motivated by consulting problems that I encountered. This book presents some of these problems and the theoretical developments motivated by them.

In the early 1950s, I assisted scientists in analyzing statistical data connected with experiments in biology and in agriculture, and I collaborated with a physicist in estimating the amount of electrical energy extracted from windmill towers placed in various parts of the country. All these data analyses were done with mechanical calculators. Computations were at that time slow and cumbersome. Due to this difficulty, models had to be developed in order to obtain mathematical solutions. Later, in the early 1960s, we could calculate with the mainframe computers, using early versions of FORTRAN and ALGOL. Data quantities were relatively small and usually the data were structured. Scientific models were very important. Results were derived mathematically, based on the models, and calculations were in many cases simple. Today, with the rapid technological developments, one could analyze big data sets with advanced software like C++, R, MINITAB, SAS, JMP, and more. Students and statisticians can perform on their laptops statistical analysis faster and more accurately than we could have done in the past. Complicated algorithms are available today for data mining, machine learning, artificial intelligence, and other facets of big data treatment. Some people claim that modeling is not necessary these days; we can rely on the available algorithms for prediction and control of systems. The book illustrates theoretical developments of statistics, motivated by consulting problems before the present era of big data sets. In that period we had to rely on modeling, sampling surveys, and simulations including boot strapping, for statistical analysis. Dynamic programming, mathematical optimization, stochastic processes, detection of change-points, and mathematics of finance were all developed during the second half of the previous century. Many of these mathematical techniques are illustrated in the chapters of this book. All these techniques continue to be important tools also in what is developed today as Data Science. Comments about training of statisticians in modern times are given in Chap. 12.

The topics treated in the book are process tracking and detection of change-points; estimation of common parameters of different distributions; survival probabilities in crossing fields with absorption points; logistics and operations research; prediction theory for finite populations; stochastic visibility in random fields; process control and reliability; random choice of fractional replications; sequential search for the maximum tolerated dose (MTD) in phase I of clinical trials; and contributions to applied probability. In Chap. 1, the events that led to studying these topics are described. The following chapters present the theoretical developments motivated by the problems discussed in the first chapter. The presentation is in the style of a seminar lecture, with references and numerical demonstrations. R programs (functions) and comments on computations are given in the Appendix.

The book is designed for the benefit of researchers, graduate students, or professional mathematical statisticians. I hope that the book will be found useful to researchers in data science as well.

McLean, VA, USA   Shelemyahu Zacks
August 2019

Acknowledgments

First of all I would like to thank Professor R.S. Kenett, who encouraged me throughout the writing of this manuscript and provided me with ideas and references. I am very grateful to Professor David Steinberg, Editor of Birkhäuser's Statistics for Industry, Technology, and Engineering (SITE) series, who read the first draft of the monograph and its revision and provided me valuable comments and suggestions. Thanks are due to the Birkhäuser Editor, Christopher Tominich and to the Assistant Editor Chris Eder, for their assistance and advice along the submission and publication process.

During my career as research statistician I collaborated and learned from Professors S.B. Littauer and S. Ehrenfeld of Columbia University; Professors H. Chernoff and H. Solomon of Stanford University; Professor W.H. Marlow of George Washington University; Professor M. Yadin of the Technion, Israel Institute of Technology; Professor R.S. Kenett of Binghamton University; Professor N. Mukhopadyay of the University of Connecticut; Professors D. Perry and S. Bar-Lev of the University of Haifa; Professor W. Stadje of the University of Osnabruck; Professor Onno Boxma of the Technical University of Eindhoven; and Professor A. DiCrescenzo of the University of Salerno. In addition I collaborated with many mathematicians, statisticians, and other scientists, as seen in the references of the present book. I hereby acknowledge their contributions. The technical assistance of Professor Yaakov Malinovsky is very much appreciated.

Last but not least, I thank my wife, Dr. Hanna Zacks, for her support and encouragement.

Contents

Abbreviations and Acronyms

ADD	Average detected delay
AMOC	At most one change
ANOVA	Analysis of variance
ARL	Average run length
BLU	Best linear unbiased
c.d.f.	Cumulative distribution function
CED	Conditional expected delay
CPW/E	Weibull/Exponential damage process
CUSUM	Cumulative sum
ECTS	Expected combined test scores
EWOC	Escalation with overdose control
FIM	Fisher information matrix
i.i.d.	Independent and identically distributed
LSE	Least squares estimator
MCCRES	Marine Corps combat readiness evaluation system
MLE	Maximum likelihood estimator
MPS	Mission performance standard
MSE	Mean squared error
MTD	Maximum tolerated dose
MTTF	Mean time to failure
MVUE	Minimum-variance unbiased estimator
OC	Operating characteristic
$p(j,m)$	Poisson p.d.f. with mean m
$P(j,m)$	Poisson c.d.f. with mean m
p.d.f.	Probability distribution function (density)
PFA	Probability of false alarm
PMLE	Pair-wise MLE
r.v.	Random variable
SHRO	Shiryaev Roberts

SPC   Statistical process control
SPRT   Sequential probability ratio test
STE   Stein-type estimator
TES   Total evaluation score

Chapter 1
Introduction

1.1 Background and Motivation

I had a long career as a research statistician, starting in the early fifties as a student and continuing till today. My first job was in 1954, at the Research Council of Israel, to assist an agronomist, Mr. N. Landau, in analyzing data on the effect of number of light hours on the yield of plants. This job continued for 2 years, doing mainly Analysis of Variance and Regression Analysis. I learned these methodologies from the books of Hald (1952), and later from the books of Snedecor and Cochran (1980) and Fisher (1925). In 1956 I assisted Dr. M. Neeman, an organic chemist, in analyzing the effects of various synergists to DDT that he developed, on the mortality of mosquitoes. For this purpose the book of Finney (1978) on the methodology of bioassays was of great help. In 1956, I worked also with a physicist, Dr. J. Frenkel, in analyzing data from wind turbines, on the amount of electric energy that can be extracted at various sites in Israel. Here we used physical models converting the wind speed to amount of electric energy. A joint paper on this subject was published in 1958, at the Journal of the Research Council of Israel. See the paper of Frenkel and Zacks (1957).

In 1958 I was hired as a statistician of the Building Research Station, at the Technion, Israel Institute of Technology, in Haifa, Israel. My task was to analyze experiments designed to test the Compressive Strength of Concrete as a function of the methods of preparation. In addition I started teaching (as an assistant) statistical methods to a class of Mechanical Engineers. In 1960 I was sent by the Technion to Columbia University in New York to study for my Ph.D. I received the Ph.D. in 1962 and was then invited to do research at the Department of Statistics, at Stanford University, as a Post-Doc research associate. I returned to the Technion in 1963 in an academic position of Senior Lecturer and worked there until 1965. From 1965 I worked at various universities in the United States, as Professor in departments of statistics or mathematical sciences until I retired in 2014. I have written close

© Springer Nature Switzerland AG 2020

S. Zacks, *The Career of a Research Statistician*, Statistics for Industry, Technology, and Engineering, https://doi.org/10.1007/978-3-030-39434-9_1

to 175 papers, over 20 chapters in edited books, and 11 books. The papers which I published dealt with the following topics:

1. Applied probability;
2. Design of experiments;
3. Sampling theory for finite populations;
4. Change-point analysis;
5. Adaptive procedures;
6. Sequential estimation and testing;
7. Distributions of stopping times;
8. Reliability analysis;
9. Inventory systems;
10. Queuing systems;
11. Stochastic visibility;
12. Actuarial science;
13. Military operations research.

The majority of the papers which I wrote were motivated by problems from physics, engineering, biostatistics, genetics, and operations research. Usually my consulting experience was limited to the Program in Logistics, at George Washington University, to the Army Operations Research group, TRASANA, in white Sands Missile Range, and a small number of industries. My main occupation was in academia, as a teaching and research professor. In that capacity 30 students received their Ph.D. degree under my guidance. I was elected Fellow of the IMS, ASA, AAAS, Member of the ISI, and in 2005 I was awarded Honorary Ph.D. from the University of Haifa. In 2012 I was awarded the degree of SUNY Distinguished Professor of Mathematical Sciences.

My objective in the present book is to illustrate theoretical contributions that evolved from consulting experience. In the Introduction (Chap. 1) these problem areas are discussed. The following chapters present the various theoretical developments associated with the problems highlighted in the Introduction, and some further contributions of other researchers on these subjects. All discussions are based on published papers. The presentation of these papers is in a lecture style. The reader will have an opportunity to see different approaches and techniques which might be helpful even in today's problem solving.

1.2 Tracking Normal Processes Subjected to Change Points

In 1962–1963 I was a Research Associate in the Department of Statistics at Stanford University, after receiving my Ph.D. at Columbia University. I did not have to teach and could devote all my time for further studies and research. My advisor there was Professor Herman Chernoff. A group of naval officers visited the department to consult about statistical problems. They raised the problem of early detection of malfunctioning of missiles, which are tested into the Pacific ocean. It was an essential security problem, in case a missile gets out of control and should be destroyed. Professor Chernoff and I formulated this as a tracking problem, in which one has to estimate the current location of the process mean as a function of time, based on sequentially obtained past observations. Simultaneously a decision is made whether the process is under control or whether it should be immediately stopped (destroyed). As a by-product, we developed a new Bayesian test for the existence of a change- point among a sample of time ordered observations. This material was published in the *Annals of Mathematical Statistics* in 1964. In addition the chapter contains material on Bayesian detection of change-points, including the Shiryaev–Roberts procedure, and material on statistical process control, including the Cumulative Sum (CUSUM) procedure.

1.3   Estimating the Common Parameters of Several Distributions

In 1964 I was a Senior Lecturer at the Technion, Israel Institute of Technology. A professor from the department of soil engineering came to my office with the following problem. They had two soil samples from different locations. They could assume that the corresponding distributions were normal with the same mean. The variances were considered different and unknown. His question was how to estimate the common mean of the two normal distributions. The problem is an interesting one since if the ratio of the variances is unknown, there exists no best unbiased estimator of the common mean. The minimal sufficient statistic is incomplete. I suggested that they would compute the weighted average of the two sample means, when the weights are the inverses of the sample variances. This was an acceptable solution at the time. I started immediately to investigate possible solutions and published three papers on this subject. Later a Ph.D. student at Tokyo University wrote his Ph.D. dissertation on this subject. A few other papers were written on this subject during the following years. In Chap. 3 my papers are discussed, as well as the contributions of other researchers.

Ten years later I started to study with my students the problem of estimating the common variance of correlated normal variables. This is an interesting problem of how to estimate the common variance in a multivariate normal distribution with arbitrary correlations. In Chap. 3 we focus attention on the case of equi-correlated normal variables having a common variance.

1.4   Survival Probabilities in Crossing Fields Having Absorption Points

In 1966 I published a paper in the *Naval Research Logistics Quarterly*, entitled "Bayes Sequential Strategies for Crossing a Field Containing Absorption Points." This paper was a result of advising a military officer, who was a master student in operations research at the Technion. It had several military applications. I continued to study similar problems for a US Army operations research group in 1975. They approached me after reading the 1966 paper, while I was working at Case Western Reserve University in Cleveland. The papers, which are discussed in Chap. 4, are focused on the following problem.

A sequential decision problem is considered in which N objects have to cross a field. Two alternative crossing paths are available. An unknown number of absorption points, J_1 and J_2, are planted at each of the crossing paths respectively. The bivariate distribution of (J_1, J_2) is given. If an object passes close to an absorption point, it may survive with probability $0 < s < 1$. If the object is absorbed, both the object and the absorption point are ruined. There is no replacement of ruined absorption points. All absorption points act independently. The objects cross the field in a consecutive order, and a crossing path can be chosen for each object. The objective is to maximize the expected number of survivors. The Bayes sequential procedure is characterized. The problem is similar to the "two-armed bandit problem," a famous problem in optimal decision making.

In 2003, I started to do research on the properties of a process called a telegraph process. This process has applications in physics, in finance in inventory theory and other areas. In 2015 I published with collaboration of Professor Antonio Di Crescenzo a paper on telegraph process subjected to an absorption point in the origin. The process hitting the origin might be absorbed with probability α or reflected back with probability $1 - \alpha$. In Sect. 4.4 the theory for computing the distribution of the time until absorption is presented. This theory has applications in finance.

1.5   Logistics and Operations Analysis for the Military

In 1968, Professor Herbert Solomon of Stanford University recommended to Dr. W.H. Marlow, the head of the Program in Logistics, at George Washington University, to invite me as a guest speaker at their seminar, and to talk with their researchers on logistics problems. The Program in Logistics at the time was supported by the Office of Naval Research, and one of the functions of the Program was to design the supply system for the Polaris submarines. After my visit I sent Dr. Marlow a memo on a two-tier system of inventory control that could be applied to their projects. I was then invited to be a consultant to the Program in Logistics, and my appointment was renewed yearly for 20 years. I was involved in many projects. In Chap. 5 three studies are presented. One is on two-tier inventory analysis for naval applications. The second one is on readiness study for the Marine Corps, which is relevant for statistical analysis of large data sets of multivariate correlated binary data. The third study is on a reliability problem of detecting the time when a system enters into a wear-out phase. This study was done for Naval Aviation.

1.6 Foundation of Sampling Surveys

In 1968, I was a professor in the mathematics and statistics department at the University of New Mexico, in Albuquerque, NM. At that time Professor Debrabata Basu was visiting the department for 2 years. Professor Basu (1969, 1971) was studying at that time the theoretical foundations of sampling surveys. We often discussed his approach, which in those days was very original, and led to the modeling approach in sampling surveys. As a result, I wrote in 1969 (see Zacks (1969b)) my first paper on the Bayesian sequential choice of units in sampling from finite populations. In the following years I published several papers on the Bayesian estimation in sampling surveys. Later in 1984, while working at Binghamton University, Professor Heleno Bolfarine, from the University of San Paulo, Brazil, came to Binghamton University as a visiting professor for 1 year. We studied together sampling surveys and published several papers and a book entitled "Prediction Theory for Finite Populations" Springer-Verlag (1992). In our approach we followed the footsteps of Professor D. Basu. This approach is discussed in Chap. 6.

1.7 Visibility in Fields with Obscuring Elements

In 1975, I was consulting to the Operations Research Group of the US ARMY. The following visibility problem was presented to me. The problem was important for visibility in the forests of Germany, where NATO forces were opposing those of the USSR. Visibility of objects in forests is lost after some distance between the observers and the targets due to randomly dispersed trees of random trunk size. Many questions may arise concerning the possible visibility of target points from a given observation point. Obviously, in the field one can either observe the targets or not. In the planning stage one has to determine the probability that a target will be visible from a given point. We need a comprehensive theory to answer such questions. We developed a stochastic visibility theory in random fields. In this theory we modeled the scattering of trees as a Poisson random field of obscuring element having random size (diameter of objects). A light ray (a laser beam) in the forest hits trees, which casts shadows on a target curve. These shadows are black cones. Different trees may cast different shadows falling on each other and creating a bigger shadow. Objects covered by shadows are unobservable from the source of light. Explicit formulas of probabilities require a great deal of geometry and trigonometry. Although the example here was on visibility in a forest, the theory can be applied to other fields, like naval navigation, astronomy, and others. I worked on these problems with Professor Micha Yadin of the Technion for almost 10 years. We published several papers on related problems. I also collected all the results in a manuscript and published in the Springer Lecture notes in Statistics (see Zacks 1994). Chapter 7 is devoted to this theory.

1.8   Sequential Testing and Estimation for Reliability
Analysis

In 1982 we opened at Binghamton University a Center for Quality Control and
Design. In this center we provided training for engineers and technician from industries in statistical quality control, reliability analysis, and design of experiments.
The type of training given is discussed in Sect. 1.12 and in Chap. 12. Another
important function of the Center was to provide consulting to industry. Several
industries around Binghamton, NY, had a problem of demonstrating to clients
the reliability of their products. For example, Universal Instruments used to build
machines which automatically place components, of different kinds, on mother
boards. Components had to fall exactly on their respective cites, without causing
shorts or other deficiencies. This is being performed very rapidly. Clients used to
send groups of engineers to observe the performance of these placement machines,
according to specified standards. The question was how long the demonstration
should last in order to obtain good reliability estimates. We introduced to this
industry the Wald sequential probability ratio test (SPRT) of acceptance or rejection.
This modification saved substantial amounts of monies to the industry. Several of the
papers discussed in Chap. 8 are focused on various sequential testing and estimation
for reliability analysis. We discuss also sequential estimation of the mean of an
exponential distribution, when the objective is to obtain estimators with bounded
risk. What are the distributions of the corresponding stopping times, and what are the
expected value and the standard deviation of the estimators of the mean at stopping.
Finally we discuss the reliability of a system which is subjected to a degradation
process in time.

1.9   Random Choice of Fractional Replication

In 1959 I met at the Technion Professor Sylvain Ehrenfeld, who came to the
Technion for 2 years as a visiting professor from Columbia University, NY.
Professor Ehrenfeld gave a course on Design of Experiments. I attended this course
and wrote my Master thesis on "Randomized Fractional Factorial Designs" under
his guidance. This was a technique to overcome possible biases in estimators of
interesting parameters under fractional factorial designs. On the basis of this thesis
we wrote together a paper which was published in the *Annals of Mathematical
Statistics* (1960). My Ph.D. dissertation at Columbia University was written as a
continuation of this paper. Several papers were later published in the *Annals of
Mathematical Statistics* on this subject. This material is discussed in Chap. 9.
     There are two approaches in the design and analysis of fractional replication.
The common approach is the so-called *conditional analysis*. The fraction of a full
factorial is chosen in such a manner that the shortest word in the group of designing
parameters is of a sufficiently large size (high resolution). This would ascertain that

the aliases of the interesting parameters represent negligible interactions, and the least squares estimators of the interesting parameters are considered to be unbiased. The other approach, which the above-mentioned papers represent, is called the *marginal approach*. Here one makes first a list of all possible fractions of the full factorial. Thus we get a population, whose units are the possible fractions. A random sample without replacement of n such units is then selected, and the observations are made on these randomly chosen fractional replications. In Chap. 9 it is proved that such random design yields unbiased estimators of the interesting parameters. In this manner we introduce an additional source of variability. This on one hand reduces bias but on the other hand increases the variance of the estimators.

1.10   Sequential Search for the Maximum Tolerated Dose in Clinical Trials

In 1971, while I was working at Case Western Reserve University in Cleveland, Professor Benjamin Eichhorn visited the department for 2 years, coming from the University of Haifa. Benjamin Eichhorn received his Ph.D. at Berkeley University, writing his dissertation on sequential search procedures. In Phase I of clinical trials, for finding the maximum tolerated dose (MTD) of a new drug, the common methodology had been the up-and-down design with a binary response. We were advised that the amount of toxicity in a patient can be determined by blood test, and the result could be considered a continuous random variable, having a log-normal distribution. The expected value of the corresponding normal distribution can be modeled as a linear function of the dosage. The coefficients of the linear regression depend on the drug and are unknown. In a sequential determination of dosages to apply to patients the coefficients of regression are estimated after each application, and estimated MTD is the estimated 0.3-quantile of the corresponding normal distribution. Concerning the variance of the normal distribution, there were two models. One model assumed that the variance is independent of the dosage. The second model assumed that the variance is proportional to the squared dosage. Eichhorn and Zacks published in the seventies several papers concerning the optimal procedures. At that time we could not interest clinicians in Cancer Phase I trials to apply our method. This was for two reasons. First, their evaluation of the amount of toxicity was different. They observed the reaction of patients to the dosages. Usually, patients who volunteered to take part in these trials were sick ones who tried already unsuccessfully other drugs in chemotherapy. The evaluation of the tolerance to toxicity was multidimensional categorical response which resulted in a score of 0–1. The score 1 was given if the toxicity was life threatening. The second reason was that the community of clinicians were used to the simple up-and-down method, which was simple and nonparametric.

In 1984 I spent a semester at the University of Sao Paulo, Brazil. After serving 3 years as chair of the department and of the center for Quality Control and Design I was eligible for semester leave with full pay, which was awarded to chair persons, in order to let them ease back into research. It was called a research semester. At the University of Sao Paulo I met a young Ph.D. student, Andre Rogatko, whose research was in human genetics and biostatistics. Several years later Dr. Rogatko came with his family to the USA and worked as a biostatistician in Fox Chase Cancer Research Center, Philadelphia. I was invited to consult at this research center on procedures for searching the Maximum Tolerated Dose (MTD) in Cancer Phase I clinical trials. Dr. Rogatko with my ex-doctoral student, James Babb, and I developed a Bayesian continuous reassessment method, with overdose control, based on categorical data. This procedure was called EWOC and was actually applied in that hospital. A special software was developed to help clinicians apply the procedure. In Chap. 10 we discuss these methods.

1.11   Contributions to Applied Probability

For 20 years since 1990 I used to visit every summer the University of Haifa to teach a graduate course in statistics. Several faculty members in the statistics department were applied probabilist. I discussed research problems with them and started to collaborate with Professors Perry, Bar-Lev, and others. Our first joint paper was published in 1999 in the journal *Stochastic Models*. The paper dealt with distributions of stopping times of compound Poisson processes with positive jumps. I continued publishing with Professor Perry, and collaborated also with Professor Wolfgang Stadje, of the University of Osnabruck, Germany, and with Professor Onno Boxma, of the Technical University of Eindhoven, Holland. We collaborated in a series of papers in queuing and inventory theories. I developed methods for analytic evaluation (in contrast to simulations) of distributions of certain functionals and their properties. These methods are presented in Chap. 11.

1.12   Current Challenges in Statistics

One can read on the INTERNET dozens of articles about the future of statistics. It is obvious that the challenges today are different from those 50 years ago. The availability of huge data sets on computers, ready for analysis, does not make the task of a statistician easier. Methods applied automatically on data may result in incomprehensible analysis. Not all the data sets are similar, and it is important that statisticians will be involved in their analysis. There is much literature on the role of statisticians in data analysis. One is referred to the papers of Breiman (2001) and Donoho (2017) and the book of Efron and Hastie (2016).

Classical statistical methodology of testing statistical hypotheses and estimating parameters of certain distributions is still valid for analyzing small size experiments which are conducted under controlled conditions. On the other hand, large data sets are generally not homogeneous and cannot be assumed as representing a simple statistical distribution. Classical methodology may not be appropriate in such cases. My approach would be to consider a large data set as a finite population and stratify it according to some reasonable criteria. Large samples can then be drawn from these strata, and suitable predictors can be evaluated for population characteristics. The validity of such predictors can be tested by cross-validation. See the book of Bolfarine and Zacks (1992), which is discussed in Chap. 6.

The question of which statistics topics to teach data science students, and how to train them, is an important issue these days. I describe below how we trained IBM engineers in the early eighties, on topics of quality control, reliability, and design of experiments.

I worked at Binghamton University from 1980 till 2014. From 1980 till 1983 I served as the Chair of the Department of Mathematical Sciences. The computers industry IBM had at that time several plants in Endicott, NY, and its vicinity. The plant in Endicott was manufacturing mid-size computers. The campus of Binghamton University (SUNY Binghamton at that time) was very close to the Endicott plant of IBM. Learning from Japanese industry, the management of IBM realized that it is important to train all their engineers and technicians in statistical quality control. The head of the quality control department at IBM Endicott addressed the university authorities asking whether the university could provide such training to their employees. The university could not provide such training within its academic framework. However, a public Center for Quality Control and Design was established. Such a center could provide lectures, training courses, consulting, and other services to the public, outside the regular academic teaching. I was asked to Chair this new center. We offered IBM to train the engineers separately from the technicians. The training of engineers to be done in workshops. Each workshop would consist of 30 engineers and last 4 weeks. Each day the participants would come to the university campus for 4 h (from 8 a.m. till 12 a.m.) and then return to their offices at IBM. In the workshop every two participants worked together on a PC (first generation of IBM PCs). We developed special software for the PCs, similar to MINITAB, which was not available in those days. In preparation of these workshops I wrote booklets for the participants, with the basic theory of statistical process control, sampling inspection schemes, and other relevant material, with many examples and exercises. The exercises were done in the classroom with the help of the instructors. Data sets were provided by IBM. At the end of the workshop, each participant received a certificate of participation, which was registered at IBM in their personal files. Technicians were trained separately to use electronic calculators, to graph data in control charts, and to perform necessary tabulations and calculations in basic data analysis. IBM accepted this proposal, and workshops lasted for close to 3 years. Several hundreds of engineers and technicians were trained. Other industries in the area, like GM, Universal Instruments, and others, adopted similar programs for their engineers and technicians. A year later we offered to IBM continuation of workshops in Design of Experiments and workshops in Reliability. These workshops were given to some engineers who took the workshop on quality control. In Chap. 12 I will summarize a textbook entitled "Modern Industrial Statistics: Design and Control of Quality and Reliability," written by R. Kenett and S. Zacks, on the basis of the workshops. Some consulting was also done to industries and papers which were written on these problems will be discussed in Chap. 8.

Chapter 2
Tracking Processes with Change Points

2.1 Tracking Procedures

In Sect. 1.2 of the Introduction we described a problem of tracking a missile in order to quickly detect significant deviation from its designed orbit. The objective is to estimate the expected value in a given time of a random process under observation, when the data available is only the past observations. The present chapter is based on the papers listed among the references. We start with the joint paper of Chernoff and Zacks (1964). For simplification, we assume that the orbit is a given horizontal line at level μ. The observations are taken at discrete time points, and they reflect the random positions of the object (missile) around μ, and the possible changes in the horizontal trend. We assume here that as long as there are no changes in the trend, the observations are realization of independent random variables, having a common normal distribution, $N(\mu, \sigma_\epsilon^2)$. Since the objective is to detect whether changes have occurred in the trend, we let $\mu_n = E\{X_n\}$, where $\{X_1, \ldots, X_n\}$ is the given data, and adopt the following model

$$X_n = \mu_1 + \sum_{i=1}^{n-1} J_{n-i} Z_{n-i} + \epsilon_n, n \geq 1, \tag{2.1}$$

where μ_n is the expected value of X_n. $J_1, J_2, \ldots$ are binary random variables, indicating whether changes in the trend occurred between the observations. The event $\{J_i = 1\}$ signifies that a change has occurred between the i-th and the $(i + 1)$st observation. Furthermore, $Z_1, Z_2, \ldots$ are random variables representing the amount of change in the trend. In addition, $\epsilon_1, \epsilon_2, \ldots$ are random variables representing observation errors. We assume that all these random variables are mutually independent. In addition, assume that $E\{\mu_1\} = \mu$, and $V\{\mu_1\} = \tau^2$. Also, $P\{J_i = 1\} = p$, for all $i = 1, 2, \ldots$ and that $E\{Z_i\} = \zeta$, $V\{Z_i\} = \sigma_Z^2$ for all $i = 1, 2, \ldots E\{\epsilon_i\} = 0, V\{\epsilon_i\} = \sigma_\epsilon^2$ for all $i = 1, 2, \ldots$. According to this

© Springer Nature Switzerland AG 2020

S. Zacks, *The Career of a Research Statistician*, Statistics for Industry, Technology, and Engineering, https://doi.org/10.1007/978-3-030-39434-9_2

simplified model, the sample path of this process is a random step function. The trend of this process is, since J and Z are independent,

$$E\{X_n\} = \mu + (n-1)p\zeta, n \geq 1,\tag{2.2}$$

and

$$V\{X_n\} = \tau^2 + (n-1)p(\sigma_Z^2 + (1-p)\varsigma^2) + \sigma_\epsilon^2, \; n \geq 1.\tag{2.3}$$

Notice that for all $1 \leq i < j \leq n$,

$$COV(X_i, X_j) = (i-1)p(\sigma_Z^2 + (1-p)\varsigma^2).\tag{2.4}$$

The trend of this model is linear, with slope coefficient $p\zeta$. The variance-covariance matrix of the vector $\mathbf{X}_5 = (X_1, \ldots, X_5)'$ is, for example,

$$V[\mathbf{X}_5] = \begin{pmatrix} v_1 & 0 & 0 & 0 & 0 \\ 0 & v_1 + v_2 & v_2 & v_2 & v_2 \\ 0 & v_2 & v_1 + 2v_2 & 2v_2 & 2v_2 \\ 0 & v_2 & 2v_2 & v_1 + 3v_2 & 3v_2 \\ 0 & v_2 & 2v_2 & 3v_2 & v_1 + 4v_2 \end{pmatrix}\tag{2.5}$$

where $v_1 = \tau^2 + \sigma_\epsilon^2$ and $v_2 = p(\sigma_z^2 + (1-p)\varsigma^2)$. Generally, the matrix $V[\mathbf{X}_n]$ is a block diagonal matrix, whose first block is in row 1 and column 1, and the second block is occupying the following $(n-1)$ rows and $(n-1)$ columns. The second block is an $(n-1) \times (n-1)$ sub-matrix, given by $v_1 I_{n-1} + v_2 K_{n-1}$, where $K_{n-1} = T_{n-1}T'_{n-1}$ and T_{n-1} is a lower triangular matrix of 1's. Thus K_{n-1} is non-singular, and $(V[\mathbf{X}_n])^{-1}$ is a corresponding block diagonal, with $1/v_1$ in the first block and $(v_1 I_{n-1} + v_2 K_{n-1})^{-1}$ in the second block. If v_1 and v_2 are known, then a linear estimator of μ_n is

$$\widehat{\mu}_n = \frac{1'_n (V[\mathbf{X}_n])^{-1} \mathbf{X}_n^*}{1'_n (V[\mathbf{X}_n])^{-1} 1_n},\tag{2.6}$$

where $\mathbf{X}_n^* = (X_n, X_{n-1}, \ldots, X_1)$. In Fig. 2.1 we see a random realization of the points X_n, and the estimator $\hat{\mu}_n$ (pointed line) of the unknown random trend μ_n. The values of $\hat{\mu}_n$ are estimated according to (2.6). The variance of $\widehat{\mu}_n$ is $1/1'_n (V[\mathbf{X}_n])^{-1} 1_n$. Thus, a possible detection of significant change in the mean μ_n, from a desired μ_0, could be using the stopping time

$$N = \min\{n \geq 1 : |\widehat{\mu}_n - \mu_0| > 2/(1'_n (V[\mathbf{X}_n])^{-1} 1_n)^{1/2}\}.\tag{2.7}$$

This detection procedure is described in Fig. 2.1 by two boundary lines around $\mu_0 = 0$. The stopping time (2.7) is the intersection of the upper boundary line and

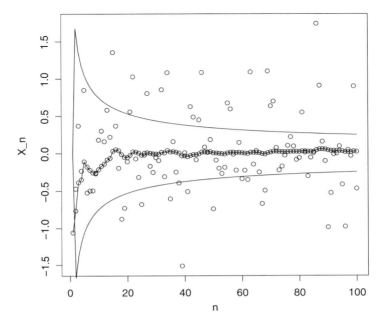

Fig. 2.1 Current means and boundary lines when $\varsigma = 0$

the graph of $\widehat{\mu}_n$. This procedure requires the prior knowledge of the values of $v1$ and $v2$.

2.2 The Bayesian AMOC Tracking Procedure

The model in the previous section allowed for possibly several changes, occurring at random times among the previously observed data. In the present section it is assumed that changes seldom occur, and there might be at most one change (AMOC) among the recent observations. The Bayesian procedure described below estimates the location of the change-point and the size of the change within a window of n_w observations. If an epoch of change is detected a window is closed, and a new window is opened. The objective is to estimate the current mean of the process.

2.2.1 The Bayesian Framework for the AMOC Procedure

The Bayesian methodology is often applied in statistical decision problems, when the optimal decision depends on unknown parameters (or components). The Bayesian procedure depends on two basic elements: the likelihood function and the information available from previous observations in a form of a prior distribution of the unknown parameter(s). The likelihood function contains all the information on the parameter(s) in the observed data. The prior distribution of the parameter(s) reflects the available information on the parameter(s) before observations commence. More specifically, let $f(x, \theta)$ denote a density function of a random variable X. The likelihood function of θ, given the observed value x, is a positive function $L(\theta, x) \propto f(x, \theta)$. The prior information on θ is provided by a density function $h(\theta)$, called a prior density. After observing x the prior density is converted to a posterior density

$$h(\theta|x) = \frac{h(\theta)L(\theta, x)}{\int_{\Theta} h(t)L(t, x)dt}. \tag{2.8}$$

The predictive distribution for future values of X is

$$p(y|x) = \int_{\Theta} f(y, \theta)h(\theta|x)d\theta. \tag{2.9}$$

We return now to the AMOC Bayesian procedure.

Let X_i have probability density $f(x, \theta_i)$, where θ_i is a parameter. The expected value is $\mu_i = E_{\theta_i}\{X_i\}$, $i = 1, \ldots, n$. The at most one change (AMOC) model assumes that $\sum_{i=1}^{n} J_i \leq 1$. That is, there is at most one change in μ among the first n observations. The location of the change is at

$$\tau_n = \min\{1 \leq i \leq n : J_i = 1\}. \tag{2.10}$$

The event $\{\tau_n = n\}$ corresponds to the case that the first change has not occurred among the first n observations. Accordingly, τ_n has the following truncated geometric prior distribution

$$P\{\tau_n = i\} = I\{i = 0\}\pi + (1 - \pi)[I\{i = 1, \ldots, n - 1\}p(1 - p)^{i-1}$$
$$+ I\{i = n\}(1 - p)^n]. \tag{2.11}$$

All observations before the change have densities with the same parameter θ, and all observations after the change have densities with a parameter ϕ different from θ. The parameters (θ, ϕ) have a joint prior distribution H, with density $h(\theta, \phi)$. The likelihood function of (τ_n, θ, ϕ), given $\mathbf{X} = \{X_1, \ldots, X_n\}$ is

$$L_n(\tau_n, \theta, \phi|\mathbf{X}) = I\{\tau_n = 0\}\prod_{i=1}^{n} f(X_i, \phi) + \sum_{j=1}^{n-1} I\{\tau_n = j\}\prod_{i=1}^{j} f(X_i, \theta)$$

$$\times \prod_{i=j+1}^{n} f(X_i, \phi) + I\{\tau_n = n\}\prod_{i=1}^{n} f(X_i, \theta). \tag{2.12}$$

According to Bayes theorem, the posterior distribution of τ_n is

$$\pi_n(i|\mathbf{X}) = P\{\tau_n = i\}\frac{\iint L_n(i, \theta, \phi|\mathbf{X})h(\theta, \phi)d\theta d\phi}{\sum_{j=1}^{n} P\{\tau_n = j\}\iint L(j, \theta, \phi|\mathbf{X})h(\theta, \phi)d\theta d\phi}. \qquad (2.13)$$

Generally, if $\{\theta_n < n\}$, then $\mu_n = \int x f(x, \phi)dx$ is a function of the parameter ϕ. Thus, a posterior distribution of μ_n, given $\mathbf{X}$, can be determined. A Bayesian estimator of μ_n can be derived for an appropriate loss function. Examples of such Bayesian estimators will be given in the following sections. If $\{\tau_n < n\}$ the data prior to the change point should not be utilized for estimating μ_n. However, since we do not know where the change point has been, we compute the posterior expected Bayesian estimator. We illustrate this in the following sections.

2.2.2 Bayesian AMOC for Poisson Random Variables

There could be many applications of this procedure in the Poisson case. For example, suppose that the number of items bought in a given period (day) has a Poisson distribution with mean λ. If the changes in the mean are rare, we apply the Bayesian AMOC procedure to track the means λ_n of consumption. In the book of Zacks (2009) the method is applied to track the mean number of days between successive mine disasters in Britain between 1851 and 1962.

Let $\{X_1, X_2, \ldots, X_n\}$ be independent Poisson random variables, with means $\mu_i, i = 1, \ldots, n$. According to the AMOC model, $\mu_i = I\{i < \tau_n\}\theta + I\{i \geq \tau_n\}\phi$. Conditional on $\{\tau_n = j\}$, where $(j = 1, \ldots, n - 1)$, $S_j = \sum_{i=1}^{j} X_i$ and $S_{n-j}^* = \sum_{i=j+1}^{n} X_i$, are minimal sufficient statistics for θ and ϕ, respectively. If $\{\tau_n = n\}$, $S_n = \sum_{i=1}^{n} X_i$ is a minimal sufficient statistic for θ. Accordingly, the joint likelihood function is

$$L_n(\tau_n, \theta, \phi|\mathbf{X}) = I\{\tau_n = 0\}e^{-n\phi}\frac{\phi^{S_n^*}}{S_n^*!} + \sum_{i=1}^{n-1} I\{\tau_n = i\}\left[\frac{\theta^{S_i}\phi^{S_{n-i}^*}}{(S_i! S_{n-i}^*!)}\right]e^{-i\theta-(n-i)\phi}$$

$$+ I\{\tau_n = n\}\frac{(\theta^{S_n}e^{-n\theta})}{S_n!} \qquad (2.14)$$

We assume that θ and ϕ are priorly independent having Gamma conjugate prior distributions with scale parameters Λ_i/ν_i and shape parameters $\nu_i, i = 1, 2$, respectively. It follows that the predictive distribution of S_j, given that $\{\tau_n = j\}, j = 1, \ldots, n$, is the negative binomial, with density

$$g(s; \psi_j, \nu_1) = \frac{\Gamma(s + \nu_1)}{s!\Gamma(\nu_1)}(1 - \psi_j)^{\nu_1}\psi_j^s, s = 0, 1, \ldots \qquad (2.15)$$

where

$$\psi_j = \frac{j \Lambda_1}{(v_1 + j \Lambda_1)}. \tag{2.16}$$

Similarly, the predictive density of S_{n-j}^* is negative binomial with parameters $\psi_{n-j}^* = (n-j)\Lambda_2/(v_2 + (n-j)\Lambda_2)$, and v_2. It follows that the posterior probability of π_n is

$$\pi_n(j|X) = [I\{1 \le j \le n-1\} p(1-p)^{j-1} g(S_j; \psi_j, v_1) g(S_{n-j}^*; \psi_{n-j}^*, v_2)$$

$$+ I\{j = n\}(1-p)^{n-1} g(S_n; \psi_n, v_1)] \left[\frac{1}{D_n(X)} \right], \tag{2.17}$$

where

$$D_n(X) = p \sum_{j=1}^{n-1} (1-p)^{j-1} g(S_j; \psi_j, v_1) g(S_{n-j}^*; \psi(n-j)^*, v_2)$$

$$+ (1-p)^{n-1} g(S_n; \psi_n, v_1). \tag{2.18}$$

An estimator of the current position of the process is

$$\hat{\mu}_n(X) = \sum_{j=1}^{n-1} \pi_n(j|X) \Lambda_2 \frac{S_{n-j}^* + v_2}{v_2 + (n-j)\Lambda_2} + \pi_n(n|X) \Lambda_1 \frac{S_n + v_1}{v_1 + n \Lambda_1}. \tag{2.19}$$

In Fig. 2.2 we illustrate the AMOC-Poisson procedure on a set of 50 observations. The first 25 are i.i.d. Poisson(5) and the next 25 are i.i.d. Poisson(15). The window size is $n_w = 5$. A detection of the change point is after the 27th observation.

2.2.3   Bayesian AMOC for Binomial Random Variables

In the present case we consider binomial data that is observed periodically. In each period we observe N_t independent trials with unknown probability of success θ_t. In Zacks (2009, pp. 270) the AMOC procedure is applied on batches of circuit boards. The binomial random variables are the number of soldering defects in a batch.

The data consists of vectors $(J_t, N_t) t = 1, 2, \ldots$, where J_t has a binomial distribution with parameters (N_t, θ_t). We discuss here a case where the number of Bernoulli trials N_t are large and a variance stabilizing angular transformation is warranted (see Johnson and Kotz 1969, p. 69). Accordingly, define

$$X_t = 2 \sin^{-1}(\sqrt{(J_t + 3/8)/(N_t + 3/4)}), t = 1, 2, \ldots \tag{2.20}$$

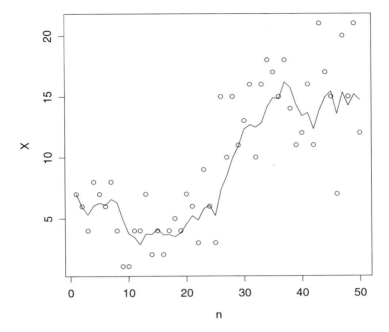

Fig. 2.2 AMOC Tracking of a Poisson current mean

The distribution of X_t is approximately normal with mean $\eta_t = 2\sin^{-1}(\sqrt{\theta_t})$ and variance $\sigma_t^2 = 1/N_t$. In the AMOC model with a window of n observations, the change point is denoted by τ_n, and

$$\eta_t = I\{t < \tau_n\}\eta_0 + I\{t \geq \tau_n\}\eta^*. \tag{2.21}$$

Define

$$N^{(j)} = \sum_{t=1}^{j} N_t,$$

$$\bar{X}_j = \frac{1}{N_j} \sum_{t=1}^{j} N_t X_t,$$

$$N_*^{(n-j)} = \sum_{t=j+1}^{n} N_t,$$

$$\bar{X}^*_{n-j} = \frac{1}{N_*^{(n-j)}} \sum_{t=j+1}^{n} N_t X_t.$$

The likelihood function of (τ_n, η_0, η^*) is equivalent to

$$
L_n(\tau_n, \eta_0, \eta^*) = \sum_{j=1}^{n-1} I\{\tau_n = j\} \exp\left\{ -\frac{1}{2} \sum_{i=1}^{j} N_i(X_i - \eta_0)^2 \right.
$$

$$
\left. -\frac{1}{2} \sum_{i=j+1}^{n} N_i(X_i - \eta*)^2 \right\}
$$

$$
+ I\{\tau_n = n\} \exp\left[-\frac{1}{2} \sum_{i=1}^{n} N_i(X_i - \eta_0)^2 \right]. \tag{2.22}
$$

This version of the likelihood function is then reduced to

$$
L_n(\tau_n, \eta_0, \eta^*) = \sum_{j=1}^{n-1} I\{\tau_n = j\} \exp\left\{ \frac{N^{(j)} N_*^{(n-j)}}{2N^{(n)}} (\bar{X}_j - \bar{X}_{n-j}^*)^2 - \frac{N^{(j)}}{2} (\bar{X}_j - \eta_0)^2 \right.
$$

$$
\left. -\frac{N_*^{(n-j)}}{2} (\bar{X}_{n-j}^* - \eta^*)^2 \right\} + I\{\tau_n = n\} \exp\left\{ -\frac{N^{(n)}}{2} (\bar{X}_n - \eta_0)^2 \right\}.
$$

We assign η_0 and η^* prior distribution of independent normal random variables with prior means ς_1 and ς_2 and prior variances σ^2 and υ^2, respectively. Let $L_n^*(\tau_n)$ denote the expected value of $L_n(\tau_n, \eta_0, \eta^*)$ with respect to the prior normal distributions of η_0 and η^*. We obtain

$$
L_n^*(\tau_n) = \sum_{j=1}^{n-1} I\{\tau_n = j\} \left(\frac{1}{(1 + N^{(j)}\sigma^2)^{1/2}(1 + N_*^{(n-j)}\upsilon^2)^{1/2}} \right)
$$

$$
\times \exp\left\{ \frac{N^{(j)} N_*^{(n-j)}}{2N^{(n)}} (\bar{X}_j - \bar{X}_{n-j}^*)^2 \right.
$$

$$
\left. -\frac{N^{(j)}}{2(1 + N^{(j)}\sigma^2)} (\bar{X}_j - \varsigma_1)^2 - \frac{N_*^{(n-j)}}{2(1 + N_*^{(n-j)}\upsilon^2)} (\bar{X}_{n-j}^* - \varsigma_2)^2 \right\}
$$

$$
+ I\{\tau_n = n\} \frac{1}{2(1 + N^{(n)}\sigma^2)^{1/2}} \exp\left\{ -\frac{N^{(n)}}{2(1 + N^{(n)}\sigma^2)} (\bar{X}_n - \varsigma_1)^2 \right\}. \tag{2.23}
$$

Accordingly, the posterior probabilities of τ_n are

$$
\pi_n(j|\mathbf{X}) = \frac{\pi_j L_n^*(j)}{\sum_{j=1}^{n} \pi_j L_n^*(j)}, \tag{2.24}
$$

where

$$\pi_j = I\{j \le n-1\} p(1-p)^{j-1} + I\{j = n\}(1-p)^n.$$

The posterior distribution of η_n given $\mathbf{X}$ is

$$G_n(\eta|\mathbf{X}) = \sum_{j=1}^{n-1} \pi_n(j|\mathbf{X}) \Phi\left(\frac{\eta - E^*_{n-j}}{D_{n-j}}\right) + \pi_n(n|\mathbf{X}) \Phi\left(\frac{\eta - E_n}{D_n}\right), \qquad (2.25)$$

where

$$E^*_{n-j} = \varsigma_2 + N^{(n-j)}_* v^2 (\bar{X}^*_{n-j} - \varsigma_2)/(1 + N^{(n-j)}_* v^2), \qquad (2.26)$$

and

$$D^2_{n-j} = v^2/(1 + N^{(n-j)}_* v^2). \qquad (2.27)$$

The Bayesian estimator of η_n given $\mathbf{X}$,

$$\eta_n = \sum_{j=1}^{n-1} \pi_n(j|\mathbf{X}) E^*_{n-j} + \pi_n(n|\mathbf{X}) E_n. \qquad (2.28)$$

2.2.4 Bayesian AMOC for Normal Random Variables

Let $X_1, X_2, \ldots, X_n$ be a sequence of independent random variables, where $X_t = \mu_t + \epsilon_t$, $t = 1, 2, \ldots, n$. We assume that the error random variables ϵ_t are distributed normally, with mean 0 and variance $\sigma^2_\epsilon = 1$ (known). The AMOC model is

$$\mu_t = I\{t \le \tau_n\}\mu_0 + I\{t > \tau_n\}(\mu_0 + Z), \qquad (2.29)$$

where τ_n, μ_0, Z are priorly independent, $\mu_0 \sim N(\mu_T, \sigma^2)$ and $Z \sim N(\delta, \tau^2)$. For $j = 1, \ldots, n-1$,

$$\bar{X}_j = \frac{1}{j} \sum_{i=1}^{j} X_i, \quad \bar{X}^*_{n-j} = \frac{1}{n-j} \sum_{i=j+1}^{n} X_i, \quad M_n = \frac{1}{n} \sum_{i=1}^{n} X_i.$$

Given $\{\tau_n = j\}$, $j = 1, \ldots, n-1$, the joint distribution of $(\mu_n, \bar{X}_j, \bar{X}^*_{n-j})$ is a trivariate normal with mean $(\mu_T + \delta, \mu_T, \mu_T + \delta)$ and covariance matrix

$$(V_j) = \begin{pmatrix} \sigma^2 + \tau^2 & \sigma^2 & \sigma^2 + \tau^2 \\ \sigma^2 & \sigma^2 + \frac{1}{j} & \sigma^2 \\ \sigma^2 + \tau^2 & \sigma^2 & \sigma^2 + \tau^2 + \frac{1}{n-j} \end{pmatrix}. \tag{2.30}$$

Thus, the posterior distribution of μ_n given $(\bar{X}_j, \bar{X}^*_{n-j})$ and $\{\tau_n = j\}$, is normal with mean $E_{j,n}$ and variance $D^2_{j,n}$. These posterior parameters are

$$E_{j,n} = w^{(1)}_{j,n}(\mu_T + \delta(1 + j\sigma^2)) + w^{(2)}_{j,n}\bar{X}_j + w^{(3)}_{j,n}\bar{X}^*_{n-j}, \tag{2.31}$$

where

$$w^{(1)}_{j,n} = [1 + n\sigma^2 + (n - j)\tau^2(1 + j\sigma^2)]^{-1}$$

$$w^{(2)}_{j,n} = j\sigma^2 w^{(1)}_{j,n}$$

$$w^{(3)}_{j,n} = 1 - w^{(1}_{j,n} - w^{(2)}_{j,n}.$$

The posterior variances are

$$D^2_{j,n} = (\sigma^2 + \tau^2 + j\sigma^2\tau^2)w^{(1)}_{j,n}. \tag{2.32}$$

The posterior distribution of μ_n given $\mathbf{X}$ and $\{\tau_n = n\}$ is normal with mean

$$E_{n,n} = \frac{\mu_T}{(1 + n\sigma^2)} + \frac{\bar{X}_n n\sigma^2}{(1 + n\sigma^2)} \tag{2.33}$$

and variance

$$D^2_{n,n} = \frac{\sigma^2}{(1 + n\sigma^2)}. \tag{2.34}$$

Finally, the posterior c.d.f of μ_n given $\mathbf{X}$ is

$$G_n(\mu|\mathbf{X}) = \sum_{j=1}^{n} \pi_n(j|\mathbf{X})\Phi\left(\frac{\mu - E_{j,n}}{D_{j,n}}\right). \tag{2.35}$$

As before,

$$\pi_n(j|\mathbf{X}) = \frac{\pi_j L^*_n(j)}{\sum_{j=1}^{n} \pi_j L^*_n(j)}, \tag{2.36}$$

and

$$L^*_n(j) = \exp\left(j\left(1 - \frac{j}{n}\right)(\bar{X}_j - \bar{X}^*_{n-j})^2\right) L^{**}_n(j), \tag{2.37}$$

where, for $j = 1, \ldots, n - 1$,

$$L_n^{**}(j) = [1 + n\sigma^2 + (n - j)\tau^2(1 + j\sigma^2)]^{-1/2}$$
$$\times \exp \left\{ -\frac{1}{2}(1 + n\sigma^2 + (n - j)\tau^2(1 + j\sigma^2))^{-1} \right.$$
$$\times \left((1 + (n - j)(\sigma^2 + \tau^2))\frac{(\bar{X}_j - \mu_T)^2}{(n - j)} - 2\sigma^2(\bar{X}_j - \mu_T) \right.$$
$$\left. \times (\bar{X}_{n-j}^* - \mu_T - \delta) + \frac{1 + j\sigma^2}{j}(\bar{X}_{n-j}^* - \mu_T\delta)^2 \right) \right\}.$$

In addition,

$$L_n^{**}(n) = \frac{1}{(1 + n\sigma^2)^{1/2}} \exp \left(-\frac{n}{2(1 + n\sigma^2)}(\bar{X}_n - \mu_T)^2 \right). \tag{2.38}$$

The Bayesian estimator of μ_n for the squared-error loss is

$$\mu_n(\mathbf{X}) = \sum_{j=1}^{n} \pi_n(j|\mathbf{X}) E_{j,n}. \tag{2.39}$$

Figure 2.3 shows the Bayesian estimator (2.38), for a sequence of normal variables, with $\mu_0 = 0$, $\tau^2 = 1$, $\sigma_\epsilon^2 = 1$, $p = 0.3$, $\varsigma = 0$ and $\sigma_Z^2 = 0.5$.

Other methods of tracking time-series, for prediction and forecasting, are given in Chapter 5 of Zacks (2009).

2.3 Sequential Detection of Change Points

In the present section we discuss detection procedures, which are not designed to estimate the current position of the process but are designed for early detection of change-points. There are two kinds of procedures for detecting change points: one is based on a given fixed sample, and the other is based on sequential procedures. A fixed sample procedure tests whether a change point exists between any two observations. There are many papers written on such fixed sample procedures. For a review including bibliography see Zacks (1983). In the present section we discuss only sequential procedures.

Sequential detection procedures are in many cases similar to tracking procedures, which have been discussed earlier. However, the objectives are different. Sequential detection of change points stops as soon as one is detected. A tracking procedure estimates the current mean and does not stop when a change-point is detected. The sequential detection procedures are especially important for statistical process control (SPC), in early detection of targets, in financial markets, etc.

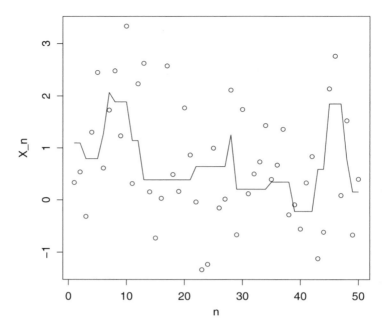

Fig. 2.3 AMOC for a normal sequence

One of the pioneering studies in this area is that of Shewhart (1931), who initiated the "3-sigma control charts." These control charts are being used widely in industrial process control all over the world. For these and similar methods in industrial quality control see Kenett and Zacks (2014, Ch. 8–10). More efficient procedures were developed by Page (1954, 1955, 1957), and are used under the name of CUSUM procedures, Bayes sequential stopping rules were developed by Shiryaev (1963), Bather (1963, 1967), Zacks and Barzily (1981), Zacks (1985), Peskir and Shiryaev (2002), and Brown and Zacks (2006).

2.3.1 Bayesian Sequential Detection, When the Distributions Before and After Are Known

We consider a case where the densities before and after the change point, i.e., $f_0(x)$ and $f_1(x)$, are known. we have to detect the change-point τ. Shiryaev (1963) considered a Bayesian procedure, where the prior distribution of τ is

$$\pi(\tau) = \pi I\{\tau = 0\} + (1 - \pi)p(1 - p)^{j-1}I\{\tau = j \geq 1\}. \tag{2.40}$$

π is the prior probability that all observations are i.i.d $f_1(x)$, while if $\{\tau = j\}$ the first j observations are i.i.d. $f_0(x)$ and the other $(n - j)$ observations are according to

$f_1(x)$. After n observations the prior distribution (2.39) is converted to the posterior distribution

$$\pi_n(\tau) = \pi_n I\{\tau = 0\} + (1 - \pi_n)p(1 - p)^{j-1} I\{\tau = j \geq 1\}, \tag{2.41}$$

where after dividing each term in the Bayes formula by $(1 - \pi) \prod_{i=1}^{n} f_0(x_i)$ we obtain

$$\pi_n = \frac{\frac{\pi}{1-\pi} \prod_{i=1}^{n} R(x_i) + p \sum_{j=0}^{n-1}(1 - p)^{j-1} \prod_{i=j+1}^{n} R(x_i)}{\frac{\pi}{1-\pi} \prod_{i=1}^{n} R(x_i) + p \sum_{j=0}^{n-1}(1 - p)^{j-1} \prod_{i=j+1}^{n} R(x_i) + (1 - p)^n}, \tag{2.42}$$

and where $R(x_i) = f_1(x_i)/f_0(x_i)$. We obtain the recursive formula, for $n = 0, 1, 2, \ldots$

$$\pi_{n+1} = \frac{(\pi_n + (1 - \pi_n)p)R(x_{n+1})}{(\pi_n + (1 - \pi_n)p)R(x_{n+1}) + (1 - \pi_n)(1 - p)}, \tag{2.43}$$

where $\pi_0 = \pi$.

Shiryaev proved that the optimal stopping time (detection time) is

$$\tau^* = \min\{n : \pi_n \geq \pi^*\}. \tag{2.44}$$

The threshold π^* is a value in $(0,1)$, which depends on p, π and the cost of observations c. For details see Zacks (2009, pp. 238–240). Simplifying (2.41) by substituting $\pi = 0$, and p close to zero, we obtain

$$\pi_n \cong \frac{\sum_{j=0}^{n-1} \Pi_{i=j+1}^{n} R(x_i)}{\sum_{j=0}^{n-1} \Pi_{i=j+1}^{n} R(x_i) + 1} = \frac{W_n}{(1 + W_n)}. \tag{2.45}$$

The statistic $W_n = \sum_{j=0}^{n-1} \Pi_{i=j+1}^{n} R(x_i)$ is called the *Shiryaev–Roberts statistic* (SHRO). This statistic can be computed recursively by the formula

$$W_{n+1} = (1 + W_n)R(x_{n+1}), n = 1, 2, \ldots W_1 = R(x_1). \tag{2.46}$$

The SHRO statistic is an asymptotically optimal statistics for detecting change-points (see Pollak (1985)). For small samples properties of this SHRO statistic, see Kenett and Zacks (2014). We can mention the following characteristics:

1. The average run length (ARL) is the number of observations until detection of change;
2. The probability of false alarm (PFA) is the probability of indicating a change before it has happened.
3. The conditional expected delay (CED) is the conditional expected number of observations, after the actual delay, until stopping.

Table 2.1 Simulation estimates of the SHRO statistics at stopping

λ_0	λ_1	τ	$E\{N\}$	$Med\{N\}$	$\max\{N\}$	$E\{W_N\}$
1	2	Inf	35.04	26	118	186.86
1	3	Inf	47.61	45	113	328.90
2	3	Inf	29.03	23	114	153.62
10	15	Inf	230.79	162	1834	253.40
10	20	Inf	493.18	322	4521	506.88
10	25	Inf	1211.94	835	8871	706.10
15	30	Inf	3121.26	2265	10,001	2827.16
17	20	15	16.51	17.50	36	33.80

Most of the sequential detection procedures stop at a finite point with probability 1. If the sequence has no change-point, i.e., $\tau = \infty$, it is desirable that ARL will be as large as possible. In this case of no change PFA $= 1$, and the CED $= 0$. Accordingly, in asymptotically optimal procedures, the characteristics of interest are the average run length, which is denoted ARL0, the PFA, and the CED. For examples of simulating these characteristics, see Kenett and Zacks (2014, p.345).

In Table 2.1 we present simulation estimates of characteristics of the SHRO for Poisson random variables. In these simulations the number of replicas is 1000. The risk factor is $\alpha = 0.01$. Thus, the critical value of the SHRO is $W* = 99$. The stopping time is denoted by N and the value of W at stopping by W_N.

We see in Table 2.1 that the SHRO is efficient under finite change point τ. When there are no change points $\tau = \infty$, the expected stopping times ARL0 are large, and the distribution of N is approximately exponential.

Recent research on the asymptotic properties of the SHRO was published by Plunchenko and Tartakovsky (2010).

2.3.2   Bayesian Detection When the Distributions Before and After Are Unknown

When the distributions before and after the change-point are unknown, or not completely known, the problem of detecting the change-point becomes more complicated. In the present section we focus attention to the case where the forms of the distributions are known, but the values of their parameters are unknown. We also assume for simplification that the change point τ is independent of the parameters. We generalize the Bayesian framework of the previous section by assuming that the parameters are random variables, with a joint density $h(\theta, \phi)$, where θ is the parameter before the change-point, and ϕ is that after the change-point. Given a sample of n observations, the predictive joint density is

$$f_H(x_1, \ldots, x_n) = \pi \int \prod_{i=1}^{n} f(x_i, \phi) h(\phi) d\phi$$

$$+ (1 - \pi) \left[p \sum_{j=1}^{n-1} (1 - p)^{j-1} \iint \prod_{i=1}^{j} f(x_i, \theta) \right.$$

$$\left. \times \prod_{i=j+1}^{n} f(x_i, \phi) h(\theta, \phi) d\theta d\phi \right]$$

$$+ (1 - p)^n \int \prod_{i=1}^{n} f(x_i, \theta) h(\theta) d\theta. \tag{2.47}$$

The posterior probability that the change has already occurred is

$$\pi_{H,n} = 1 - \frac{(1 - p)^n \int \prod_{i=1}^{n} f(x_i, \theta) h(\theta) d\theta}{f_H(x_1, \ldots, x_n)}. \tag{2.48}$$

We stop the search when $\pi_{H,n}$ is greater than a threshold smaller than 1. We show later an optimal stopping obtained by Dynamic Programming. We continue by applying the above Bayesian detection of a change in the success probability of a sequence of Bernoulli trials. This example is taken from the paper of Zacks and Barzily (1981).

Let $J_1, J_2, \ldots$ be a sequence of binary independent random variables with $P\{J_n = 1\} = w_n$, $n = 1, 2, \ldots$. We denote by $\mathbf{X}_n$ the vector of the first n observations. These random variables are conditionally independent, given the change-point τ. We assume that

$$w_n = I\{n < \tau\}\theta + I\{n \geq \tau\}\phi, \text{ where } 0 < \theta < \phi < 1. \tag{2.49}$$

We consider a uniform prior distribution for (θ, ϕ) over the simplex $\Omega = \{(\theta, \phi) : 0 < \theta < \phi < 1\}$. The minimal sufficient statistics, given that $\tau = j$, are $T_j = \sum_{i=1}^{j} J_i$, and $T_{n-j}^{(n)} = \sum_{i=j+1}^{n} J_i$. Thus, substituting in (2.46) and (2.47) the probability functions of the binomial distributions of T_j and $T_{n-j}^{(n)}$ and integrating (θ, ϕ) over the simplex Ω, we obtain

$$\pi_n(\mathbf{X}_n) = 1 - \frac{(1 - p)^n B(T_n + 1, n - T_n + 2)}{D_n(\mathbf{X}_n)}, \tag{2.50}$$

where

$$D_n(\mathbf{X}_n) = \frac{\pi}{(1-\pi)} B(T_n + 2, n - T_n + 1)$$

$$= p \sum_{j=1}^{n-1} (1-p)^{j-1} B(T_{n-j}^{(n)} + 1, n - j - T_{n-j}^{(n)} + 1)$$

$$\times \sum_{i=0}^{T_{n-j}^{(n)}} \binom{n-j-1}{i} B(T_j + i + 1, n - i - T_j + 2)$$

$$+ (1-p)^n B(T_n + 1, n - T_n + 2). \tag{2.51}$$

In these formulas $B(a, b) = \int_0^1 x^{a-1}(1-x)^{b-1} dx$ is the complete beta function. As we mentioned before, a possible stopping rule is to stop as soon as $\pi_n(\mathbf{X}_n) \geq \gamma$, where $0 < \gamma < 1$. This rule is not necessarily optimal. We describe now the truncated Dynamic Programming method, which yields optimal stopping rule, but is much more difficult to execute.

Let the penalty for stopping before the change-point be 1 (loss unit). Let c be the penalty for each time unit of delay after the change-point. Consider a truncated procedure that must stop after a specified number of observations, say N^* if it has not stopped before. Let $R_n^{(j)}(\mathbf{X}_n)$ denote the minimal posterior risk after n observations, if at most j more observations are allowed. If $j = 0$,

$$R_n^{(0)}(\mathbf{X}_n) = 1 - \pi_n(\mathbf{X}_n). \tag{2.52}$$

If $j = 1$, the anticipated risk is,

$$\bar{R}_n^{(1)}(\mathbf{X}_n) = c\pi_n(\mathbf{X}_n) + 1 - E\{\pi_{n+1}(\mathbf{X}_{n+1}) | \mathbf{X}_n\}. \tag{2.53}$$

The associated minimal risk is

$$R_n^{(1)}(\mathbf{X}_n) = \min\{R_n^{(0)}(\mathbf{X}_n), \bar{R}_n^{(1)}(\mathbf{X}_n)\}$$
$$= 1 - \pi_n(\mathbf{X}_n) + \min\{0, c\pi_n(\mathbf{X}_n) - p(1 - \pi_n(\mathbf{X}_n))\}. \tag{2.54}$$

Let $\pi^* = p/(c + p)$. It follows from (2.53) that if $\pi_n(\mathbf{X}_n) < \pi^*$ when $j = 1$ it is optimal to stop. Define recursively, for $j = 1, 2, \ldots$

$$R_n^{(j)}(\mathbf{X}_n) = 1 - \pi_n(\mathbf{X}_n) + \min\{0, c\pi_n(\mathbf{X}_n) - p(1 - \pi_n(\mathbf{X}_n)) + M_n^{(j-1)}(\mathbf{X}_n)\}, \tag{2.55}$$

where

$$M_n^{(j)}(\mathbf{X}_n) = E\{\min\{0, c\pi_{n+1}(\mathbf{X}_{n+1}) - p(1 - \pi_{n+1}(\mathbf{X}_{n+1})) + M_{n+1}^{(j-1)}(\mathbf{X}_{n+1})\} | \mathbf{X}_n\}. \tag{2.56}$$

One can prove by induction on n that

$$M_n^{(j)}(\mathbf{X}_n) < M_n^{(j-1)}(\mathbf{X}_n), \text{ a.s. for all } j = 1, 2, \ldots \text{ where } M_n^{(0)}(\mathbf{X}_n) = 0, \text{ for all n.} \tag{2.57}$$

Define the sequence of boundaries,

$$b_n^{(j)}(\mathbf{X}_n) = \min\left\{ \pi^* - \frac{M_n^{(j-1)}(\mathbf{X}_n)}{(c+p)}, 1 \right\}, j = 1, 2, \ldots \tag{2.58}$$

From (2.56) it follows that

$$b_n^{(j)}(\mathbf{X}_n) \geq b_n^{(j-1)}(\mathbf{X}_n) \text{ a.s. for all } j = 1, 2, \ldots \text{ and all n .} \tag{2.59}$$

It follows for the truncated case that the boundary $b_n^{(N^*)}(\mathbf{X}_n)$ is the optimal boundary to stop when $\pi_n(\mathbf{X}_n)$ exceeds it. It is generally difficult to obtain an explicit formula for the optimal boundary. However, for large values of n, the explicit boundary $b_n^{(2)}(\mathbf{X}_n)$ coincides with the optimal one. Notice that

$$b_n^{(2)}(\mathbf{X}_n) = \min\left\{ \pi^* - \frac{M_n^{(1)}(\mathbf{X}_n)}{(c+p)}, 1 \right\}, \tag{2.60}$$

where

$$M_n^{(1)}(\mathbf{X}_n) = E\{\min\{0, c\pi_{n+1}(\mathbf{X}_{n+1}) - p(1 - \pi_{n+1}(\mathbf{X}_{n+1}))\}|\mathbf{X}_n\}.$$

Zacks and Barzily (1981) derived an explicit function for $M_n^{(1)}(\mathbf{X}_n)$ in terms of $D_n(\mathbf{X}_n)$, which is defined in (2.50). This is

$$M_n^{(1)}(\mathbf{X}_n) = [cD_{n+1}(\mathbf{X}_n, 0)/D_n(\mathbf{X}_n) - (c+p)(1-p)(1 - \pi_n(\mathbf{X}_n))$$
$$\times (n + 2 - T_n)/(n+3)]^-. \tag{2.61}$$

From Eq. (2.49) it follows immediately that $\lim_{n \to \infty} \pi_n(\mathbf{X}_n) = 1$. It follows that $\lim_{n \to \infty} b_n^{(2)}(\mathbf{X}_n) = \pi^*$.

In Table 2.2 we present three simulation runs to illustrate the stopping rule

$$n = \min\{n > 0 : \pi_n(\mathbf{X}_n) \geq b_n^{(2)}(\mathbf{X}_n)\}. \tag{2.62}$$

In Table 2.2 we present the values of T_n, π_n, and $b_n^{(n)}$ with the following parameters: $\pi = 0, p = 0.01, c = 0.06, \theta = 0.30, \phi = 0.7$, and the change point is at $\tau = 11$. In this case $\pi^* = 0.14286$.

We see in Table 2.2 that in Run 1, the stopping time is at $N_1 = 18$, and the boundary reached the limiting value π^* at $n^* = 26$. In Run 2 we have $N_2 = 12$ and $n^* = 14$. In Run 3 $N_3 = 19$ and $n^* = 20$.

Table 2.2 Posterior probabilities and boundary values

n	T_n	π_n	$b_n^{(2)}$	T_n	π_n	$b_n^{(2)}$	T_n	π_n	$b_n^{(2)}$
Run	1	1	1	2	2	2	3	3	3
2	1	0.01502	0.41039	0	0.00838	0.44057	0	0.00838	0.44057
3	2	0.03919	0.35260	0	0.01091	0.40993	0	0.01091	0.40993
4	2	0.02829	0.34243	0	0.01294	0.38782	1	0.05828	0.34610
5	3	0.05461	0.30473	0	0.01464	0.37105	2	0.13568	0.28698
6	4	0.08681	0.27341	0	0.01612	0.35787	2	0.07454	0.29882
7	4	0.05783	0.27754	0	0.01741	0.34721	3	0.13648	0.25537
8	4	0.04712	0.27790	0	0.01857	0.33839	3	0.09103	0.26649
9	5	0.07394	0.25505	1	0.10809	0.27296	3	0.07097	0.27084
10	5	0.05786	0.25820	1	0.06716	0.28453	3	0.06009	0.27247
11	5	0.05054	0.25939	2	0.18662	0.21179	3	0.05349	0.27283
12	6	0.07775	0.24003	3	0.38013	0.14286	4	0.09568	0.24351
13	7	0.11325	0.22092	3	0.24004	0.16652	4	0.07539	0.24913
14	8	0.15840	0.20194	4	0.40718	0.14286	5	0.12311	0.22068
15	9	0.17397	0.18322	5	0.58537	0.14286	5	0.09587	0.22919
16	9	0.13504	0.19864	6	0.73529	0.14286	6	0.14641	0.20182
17	10	0.17519	0.18267	7	0.84049	0.14286	6	0.11501	0.21205
18	11	0.22231	0.16691	8	0.90617	0.14286	7	0.16692	0.18577
19	12	0.27619	0.15158	8	0.80757	0.14286	8	0.23587	0.15732
20	12	0.19094	0.16777	9	0.87581	0.14286	9	0.32355	0.14286
21	13	0.23261	0.15415	10	0.92056	0.14286	10	0.42758	0.14286
22	14	0.27920	0.14286	11	0.94911	0.14286	11	0.29560	0.14286
23	14	0.20873	0.15506	12	0.96716	0.14286	11	0.37500	0.14286
24	14	0.17030	0.15506	12	0.96716	0.14286	11	0.37500	0.14286
25	15	0.20286	0.15346	13	0.98586	0.14286	12	0.35599	0.14286
26	16	0.23967	0.14286	14	0.96855	0.14286	12	0.28599	0.14286
27	17	0.28088	0.14286	14	0.94183	0.14286	13	0.34813	0.14286
28	18	0.32644	0.14286						
29	19	0.37604	0.14286						

2.4 Statistical Process Control

The statistical theory of process control was initiated in 1924 by Walter A. Shewhart at Bell Telephone Laboratories, with the first design of the control chart for the mean. Shewhart's book (1931) gave the theory wide publicity. His colleagues H.F. Dodge and H.G. Romig applied statistical sampling theory to the other branch of statistical quality control, namely, *Sampling Inspection Schemes*. Many books and articles were written following these pioneering works on the subject of statistical quality control. Statistical process control (SPC) deals with the early detection of changes at the process level, which occur at unknown time points.

2.4.1 The Cumulative Sum Procedure

The cumulative sums (CUSUM) procedure was introduced by Page (1954, 1955, 1957) for detecting a change point in the level of a process. The CUSUM procedure was introduced in order to increase the efficiency of Shewhart 3-sigma control charts. We start with a simple demonstration of Page's procedure.

Let $X_t, t = 1, 2, \ldots$ be a sequence of independent random variables, having normal distributions with means $\mu_t = I\{t \le \tau\}\mu + I\{t > \tau\}(\mu + \Delta)$, and common standard deviation $\sigma_t = 1$. We also assume that $\Delta > 0$. As before, τ denotes the change-point. Let (K, h) be positive constants, and consider the sequence of nonnegative sums

$$S_t = \max\{S_{t-1} + (X_t - K)^+, 0\}, t = 1, 2, \ldots \quad \text{and} \quad S_0 = 0. \quad (2.63)$$

The idea is to add to the previous sum the difference between the new value of X_t and the positive constant K, only if the difference is positive. Stopping occurs when $S_t > h$. Such a sequence may stay for a while on the same level, and then have a positive jump. The length of cycles at which the sequence stays at the same level before the change-point are independent and identically distributed, and are called *renewal cycles*. After the change-point the probability of staying at the same level is smaller than before (it depends on the magnitude of Δ, and we expect that the sequence of cumulative sums (CUSUM) will cross the threshold h fast). The constants K and h are determined on the basis of Wald sequential probability ratio test procedure (SPRT), which is described below.

Let $X_t, t = 1, 2, \ldots$ be a sequence of independent random variables having a common distribution with density $f(x, \theta)$. In order to test the hypothesis $H_0 : \theta < \theta_0$, against the alternative $H_1 : \theta > \theta_1$, where $\theta_0 < \theta_1$, with error probabilities α and β, the most efficient test procedure is the Wald SPRT. This is a sequential procedure based on the cumulative sums

$$S_t = \sum_{i=1}^{t} \log(R(X_i)), \quad (2.64)$$

where $R(X_i) = f(X_i, \theta_1)/f(X_i, \theta_0)$. We set two boundaries $B_L = \log(\beta/(1 - \alpha))$ and $B_U = \log((1-\beta)/\alpha)$. As long as $B_L < S_t < B_U$ we add additional observation. As soon as $S_t < B_L$ we stop and accept the null hypothesis. If on the other hand $S_t > B_U$ we stop and accept the alternative hypothesis. For our detection procedure with upwards shift, we need only an upper boundary, with $\beta = 0$. This yields $B_U = -\log(\alpha)$. The lower boundary vanishes, since $B_L = -\infty$.

The following are three examples for determining the constants (K, h) for the CUSUM.

1. *Normal Distribution*
   The likelihood ratio is

$$log(R(X)) = (\theta_1 - \theta_0)(X - (\theta_1 + \theta_0)/2)/\sigma^2. \tag{2.65}$$

Hence, $K = (\theta_1 + \theta_0)/2$, and $h = -\sigma^2 \log(\alpha)/(\theta_1 - \theta_0)$.

2. *Binomial Distribution*

$$K = n \log(1 - \theta_0)/(1 - \theta_1)/\log(\theta_1/(1 - \theta_1)/\theta_0(1 - \theta_0)) \tag{2.66}$$

and

$$h = -\log(\alpha)/\log(\theta_1/(1 - \theta_1)/\theta_0(1 - \theta_0)) \tag{2.67}$$

3. *Poisson Distribution*

$$K = (\lambda_1 - \lambda_0)/\log(\lambda_1/\lambda_0), \tag{2.68}$$

and

$$h = -\log(\alpha)/\log(\lambda_1/\lambda_0). \tag{2.69}$$

In Table 2.3 we present simulation estimates of the characteristics $E\{N\}$, PFA, CED of the CUSUM procedure for a sequence of Poisson random variables. At the change-point τ the mean λ_0 changes to λ_1. The risk level is $\alpha = 0.01$, and the number of simulation runs is 1000.

We see in Table 2.3 that the expected stopping time is very close to the actual place of the change-point. The probability of false alarm, PFA, is relatively small. The conditional expected delay is between one to two observations. Since the actual value of τ is unknown, the suggestion in various papers is to characterize the procedure by the expected value of τ CED, with respect to the geometric distribution of τ. This index is called Average Detection Delay, or ADD, i.e.,

$$ADD = p \sum_{j=1}^{\infty} (1 - p)^{j-1} E\{N - \tau | N > \tau\}\tau. \tag{2.70}$$

Table 2.3 Characteristics of CUSUM for Poisson random variables

λ_0	λ_1	τ	$E\{N\}$	PFA	CED
10	20	15	16.284	0.063	1.652
10	20	20	20.727	0.105	1.573
10	30	15	15.970	0.010	1.069
10	30	25	25.989	0.007	1.075
10	30	30	30.913	0.012	1.056
10	30	40	40.645	0.023	1.063

Table 2.4 Simulated
estimates of ADD, for
CUSUM with $\alpha = 0.01$

λ_0	λ_1	ADD
10	15	3.528
10	20	1.746
10	25	1.244
10	30	1.081
10	35	1.019

In Table 2.4 we present simulated values of ADD, for $p = 0.1$, while the CED is simulated as in Table 2.3. The R program for these calculations is given in the Appendix.

There are many papers written on the properties of the CUSUM procedures. Lorden (1971) proved that the CUSUM procedure is asymptotically minimax. Zacks (2004) derived the exact distribution of (2.61), when the distribution of X is a continuous time Poisson process.

Chapter 3
Estimating the Common Parameter
of Several Distributions

As described in Sect. 1.3, the consulting problem was that of estimating the common mean of two normal distributions, when the data of the two samples are given, and the corresponding variances are unknown. There are two types of problems. One is an estimation problem, when the samples from the two distributions are given. In this case we are in the estimation phase. Prior to this phase there is a design phase, in which one has to determine how many observations to draw from each distribution. Obviously, if we know which distribution has the smaller variance, then the best is to draw all the observations from that distribution. If this information is not available, the design problem becomes very difficult if the objective is to minimize the total number of observations, under the constraint that the variance (or mean-squared error) of the estimator satisfies certain condition. This design problem is similar to the well-known "two-armed bandit problem." The present chapter is devoted to the estimation problem, when two or more samples of equal size are given from the corresponding distributions. For a sequential design problem, when the value of one variance is known, see the paper of Zacks (1974).

In this chapter we start with the contributions of Zacks (1966b, 1970b) to the theory about this problem, and proceed by short summary of the ensuing contributions of Brown and Cohen (1974), Cohen and Sackrowitz (1974), Kubokawa (1987a), Keller and Olkin (2004), and George (1991). We conclude the chapter with the problem of estimating the common variance of correlated normal distributions, as in the papers of Zacks and Ramig (1987), Ghezzi and Zacks (2005), and Hanner and Zacks (2013). Notice that if the two samples are mutually independent the common variance is estimated by the usual procedure of pooling the within samples variances, as in Analysis of Variance (ANOVA). On the other hand, if the variables in the samples are correlated, there is a loss of information due to the correlations, and one needs a more sophisticated procedure for estimating the common variance. We restrict attention to the case of equicorrelated samples.

© Springer Nature Switzerland AG 2020
S. Zacks, *The Career of a Research Statistician*, Statistics for Industry,
Technology, and Engineering, https://doi.org/10.1007/978-3-030-39434-9_3

3.1 Estimating the Common Mean of Two Normal Distributions with Samples of Equal Size

Let $\{X_1, X_2, \ldots, X_n\}$ be i.i.d. normal $N(\mu, \sigma_x^2)$ random variables (a random sample), and $\{Y_1, Y_2, \ldots Y_n\}$ i.i.d. normal $N(\mu, \sigma_y^2)$ random variables. The problem is to estimate the common mean μ of the two normal distributions. If the two variances σ_x^2 and σ_y^2 are known, then the minimal sufficient statistics are the corresponding sample means $\bar{X}_n$ and $\bar{Y}_n$. Each one is an unbiased estimator of the common mean μ. Combining the information from the two samples, the best linear unbiased estimator of μ is

$$T_n = \frac{\rho}{1+\rho}\bar{X}_n + \frac{1}{1+\rho}\bar{Y}_n = \frac{1}{1+\omega}\bar{X}_n + \frac{\omega}{1+w}\bar{Y}_n, \tag{3.1}$$

where $\rho = \sigma_y^2/\sigma_x^2$ and $\omega = \sigma_x^2/\sigma_y^2$. The variance of this estimator is

$$V\{T_n\} = \frac{\sigma_x^2\rho}{n(1+\rho)} = \frac{\sigma_y^2\omega}{n(1+\omega)}. \tag{3.2}$$

It follows that $V\{T_n\} < \min\{V\{\bar{X}_n\}, V\{\bar{Y}_n\}\}$. When the variances are unknown the common approach was to substitute for ρ the maximum-likelihood (MLE) estimator $R = Q_{y,n}/Q_{x,n}$, where $Q_{x,n} = \sum_{i=1}^n (X_i - \bar{X}_n)^2$ and $Q_{y,n} = \sum_{i=1}^n (Y_i - \bar{Y}_n)^2$. Notice that the minimal sufficient statistics in this case are the sample means and the sample sum of squares of deviations, i.e., $\{\bar{X}_n, \bar{Y}_n, Q_{x,n}, Q_{y,n}\}$. These statistics are, however, incomplete, since $E\{\bar{X}_n - \bar{Y}_n\} = 0$ while $P\{\bar{X}_n - \bar{Y}_n = 0\} = 0$. Therefore, there is no unique best unbiased estimator of the common mean μ. This is the reason for the many papers which were published on this problem.

Consider the simple estimator

$$\hat{T}_n = \frac{R}{1+R}\bar{X}_n + \frac{1}{1+R}\bar{Y}_n. \tag{3.3}$$

This estimator is unbiased, since $\bar{X}n$ and $\bar{Y}_n$ are independent of R. The variance of this estimator is

$$V\{\hat{T}_n\} = \frac{\sigma_x^2}{n}\left(E\left\{\left(\frac{R}{1+R}\right)^2\right\} + \rho E\left\{\left(\frac{1}{1+R}\right)^2\right\}\right). \tag{3.4}$$

Recall that R is distributed like $\rho F[n-1, n-1]$, where $F[n-1, n-1]$ is the variance ratio statistic, having the F-distribution, with $n-1$ and $n-1$ degrees of freedom. Accordingly, the p.d.f. of R is

$$f_R(x; \rho, n) = \frac{1}{\rho}\left(\frac{x}{\rho}\right)^{(n-3)/2} / \left(B\left(\frac{n-1}{2}, \frac{n-1}{2}\right)\left(1+\frac{x}{\rho}\right)^{n-1}\right), \tag{3.5}$$

Table 3.1 Variance of $\hat{T}_n$, with $\rho = 2$

n	$V\{\hat{T}_n\}$
10	0.05884
15	0.03845
20	0.02856
30	0.01886
50	0.01223
100	0.00559

where $B(\frac{n-1}{2}, \frac{n-1}{2}) = \int_0^1 x^{(n-3)/2}(1-x)^{(n-3)/2}dx$ is the complete beta function. Hence,

$$E_\rho\left\{\left(\frac{R}{1+R}\right)^2\right\} = \int_0^\infty \left(\frac{x}{1+x}\right)^2 f_R(x; \rho, n)dx. \tag{3.6}$$

Similarly,

$$E_\rho\left\{\left(\frac{1}{1+R}\right)^2\right\} = \int_0^\infty \left(\frac{1}{1+x}\right)^2 f_R(x; \rho, n)dx. \tag{3.7}$$

These integrals can be computed numerically (see R-software) with a high degree of accuracy. In the case that $\rho = 1$, the variance (3.4) is equal to $\sigma_x^2/2n$. In Table 3.1 we present the values of the variance (3.4) with $\sigma_x^2 = 1$, and value of $\rho = 2$, when $n = 10, 15, 20, 30, 50, 100$.

Since $1/F[n, n] =_{dis} F[n, n]$, the variance function (3.4) satisfies $V_\rho = V_{1/\rho}$. This is an inverse symmetry around $\rho = 1$. We see also another property, namely $V_\rho\{\hat{T}_n\} \leq \min\{V\{\bar{X}_n\}, V\{\bar{Y}_n\}\}$. This property was proved first by Graybill and Weeks (1959), for all $n > 9$, by combining inter-block information in balanced incomplete blocks. See also Kubokawa (1987a,b).

3.1.1 Bayesian Estimator

Another approach is to derive a Bayesian estimator for the weight function $\varsigma = \rho/(1+\rho)$. The Bayesian estimator with respect to a squared-error loss function is the expected value of the posterior distribution of ς. We consider prior distribution for ρ, with c.d.f. $H(\rho) = \rho/(1+\rho)$ for $\rho \geq 0$. Let $R = Q_y^2/Q_x^2$ and $W = (\bar{Y}_n - \bar{X}_n)^2$. The distributions of these statistics depend on ρ. Therefore, the posterior density of ρ is

$$h_n(\rho|R, W) = \frac{D(R, W)[\rho^{(n-1)/2}\exp\{-\frac{W}{2(1+\rho)}\}}{((1+\rho)^{5/2}(R+\rho)^{n-1})]}, \tag{3.8}$$

where

$$D(R, W) = \cfrac{1}{\int_0^\infty \rho^{n-1/2} \cfrac{\exp\left\{-\frac{W}{2(1+\rho)}\right\}}{(1+\rho)^{5/2}(R+\rho)^{n-1}d\rho}}. \tag{3.9}$$

By making the transformation $z = \rho/(1 + \rho)$ we obtain

$$h_n(z|R, W) \propto \frac{z^{(n-1)/2}(1-z)^{n/2} \exp \frac{-W(1-z)}{2}}{(z + R(1-z)^{n-1})}. \tag{3.10}$$

Thus, let $\varsigma = \rho/(1 + \rho)$, and let $\varsigma_B(R, W)$ denote the Bayesian estimator of ς for the squared-error loss, then

$$\varsigma_B(R, W) = E\left\{\frac{\rho}{(1 + \rho)}|R, W\right\} = \int_0^1 zh_n(z|R, W)dz. \tag{3.11}$$

In Table 3.2 we present several values of this Bayesian estimator, for randomly chosen values of R and W.

We see that the Bayesian estimator yields values close to the true ones. The estimator of the common mean, based on the Bayesian estimator of ς, is

$$\hat{\mu}_B = \varsigma_B(R, W)\bar{X}_n + (1 - \varsigma_B(R, W))\bar{Y}_n = \bar{Y}_n + \varsigma_B(R, W))(\bar{X}_n - \bar{Y}_n). \tag{3.12}$$

In Table 3.3 we show simulation estimates of the variance $V\{\hat{\mu}_B\} = E\{(\hat{\mu}_B - \mu)^2\}$. In this simulation we let $\mu = 0$, $\sigma_x^2 = 1$. The number of replicas is $N_s = 1000$.

We see in Table 3.3 that the variance of $\hat{\mu}_B$ is significantly smaller than that of the variance of $\hat{T}_n$, as shown in Table 3.1.

Table 3.2 Bayesian estimators $\varsigma_B(R, W)$

n	ρ	ς	$\varsigma_B(R, W)$
10	2	0.6667	0.6391
20	2	0.6667	0.6578
50	2	0.6667	0.6706
10	3	0.7500	0.7134
50	3	0.7500	0.7772

Table 3.3 Simulation estimates of $V\{\hat{\mu}_B\}$

ρ	n	$V\{\hat{\mu}_B\}$
2	10	0.068911
2	20	0.036560
2	30	0.023670
2	50	0.013387
2	100	0.006980

3.1.2 Pretest Estimator

A pretest estimator is an estimator which follows the result of hypothesis testing. In the present context, the hypothesis for test is $H_0 : \rho = 1$, against the alternative $H_1 : \rho \neq 1$. If H_0 is accepted we estimate the common mean by the grand average $\hat{\mu}_G = (\bar{X}_n + \bar{Y}_n)/2$; else, if H_0 is rejected, we estimate the common mean by $\hat{T}_n$. The efficiency of this pretest estimator was studied in Zacks (1966a). We consider here the common F-test of H_0, namely, accept the null hypothesis at level of significance α, if $I\{\frac{1}{F_{1-\alpha/2}} < R < F_{1-\alpha/2}\} = 1$, where F_p is the p-th quantile of $F[n-1, n-1]$. Accordingly, the pretest estimator of the common mean μ is

$$\hat{\mu}_{PT} = I\left\{\frac{1}{\rho^*} < R < \rho^*\right\}\hat{\mu}_G + \left(1 - I\left\{\frac{1}{\rho^*} < R < \rho^*\right\}\right)\hat{T}_n, \qquad (3.13)$$

where $\rho^* = F_{1-\alpha/2}$. Furthermore, $\hat{\mu}_{PT}$ is unbiased and the conditional variance of this estimator, given R, is

$$V\{\hat{\mu}_{PT}|R\} = I\left\{\frac{1}{\rho^*} < R < \rho^*\right\}\frac{1+\rho}{4n}$$
$$+ \left(1 - I\left\{\frac{1}{\rho^*} < R < \rho^*\right\}\right)\frac{1}{n}\frac{R^2+\rho}{(1+R)^2}. \qquad (3.14)$$

Indeed, $I^2 = I$, and $(1 - I)^2 = 1 - I$. It follows that

$$V\{\hat{\mu}_{PT}\} = E\{V\{\hat{\mu}_{PT}|R\}\}$$
$$= \frac{1+\rho}{4n}\int_{1/\rho^*}^{\rho^*} f_R(x; \rho, n)dx$$
$$+ \frac{1}{n}\left(\int_0^{1/\rho^*} + \int_{\rho^*}^{\infty}\right)f_R(x; \rho, n)\frac{x^2+\rho}{(1+x)^2}dx. \qquad (3.15)$$

In Table 3.4 we present a few values of this variance, with $\rho^* = F_{.975}[n-1, n-1]$, and $1/\rho^* = F_{.025}[n-1, n-1]$.

Comparing Table 3.4 with Table 3.1 we realize, as expected, that the estimator $\hat{T}_n$ is more efficient than the pretest estimator $\hat{\mu}_{PT}$.

Table 3.4 Variance of $\hat{\mu}_{PT}$

n	ρ	$V\{\hat{\mu}_{PT}\}$
10	2	0.075502
20	2	0.036802
50	2	0.013989
10	3	0.093513
20	3	0.038410
50	3	0.013989

3.2 Bayes Equivariant Estimators

All the estimators discussed above are equivariant with respect to the group of location and scale transformations. For the definition and properties of equivariant estimators, see Zacks (2014, pp. 343). Briefly explaining, let $\mathbf{X} = (X_1, \ldots, X_n)'$ and $\mathbf{Y} = (Y_1 \ldots, Y_n)'$ be vectors of observations, subjected to the same transformation $\mathbf{X} \mapsto a1 + b\mathbf{X}$, where $b > 0$, then an estimator $\hat{\mu}_n(\mathbf{X}, \mathbf{Y})$ is equivariant if $\hat{\mu}_n(a1 + b\mathbf{X}, a1 + b\mathbf{Y}) = a + b\hat{\mu}_n(\mathbf{X}, \mathbf{Y})$. Accordingly, an equivariant estimator of the common mean is of the form

$$\hat{\mu}_\Psi(\bar{X}_n, \bar{Y}_n, Q_{x,n}, Q_{y,n}) = \bar{X}_n + (\bar{Y}_n - \bar{X}_n)\Psi(Z_{x,n}, Z_{y,n}), \qquad (3.16)$$

where Ψ is a function of the maximal invariant statistics

$$Z_{x,n} = \frac{Q_{x,n}}{(\bar{Y}_n - \bar{X}_n)^2}, \ Z_{y,n} = \frac{Q_{y,n}}{(\bar{Y}_n - \bar{X}_n)^2}. \qquad (3.17)$$

Let $\hat{\mu}_\Psi$ denote such an equivariant estimator. Equation (3.17) can be written as

$$\hat{\mu}_{PT} = (1 - \Psi)\bar{X}_n + \Psi\bar{Y}_n.$$

As proved in Zacks (1970a), every such equivariant estimator is an unbiased estimator of μ. We develop now a formula for the best equivariant estimator under the invariant quadratic loss function

$$L(\hat{\mu}_\Psi, \mu) = \frac{(\hat{\mu}_\Psi - \mu)^2}{\sigma_x^2}. \qquad (3.18)$$

The parameter space of such estimators is

$$\Theta = \{(\mu, \sigma, \rho) : -\infty < \mu < \infty; \ 0 < \sigma < \infty; \ 0 < \rho < \infty\}. \qquad (3.19)$$

The orbits of the location and scale transformations are all the subspaces with a fixed $\rho > 0$, since the parameter ρ is invariant with respect to these transformations, i.e.,

$$\Theta_\rho = ((\mu, \sigma, \rho) : -\infty < \mu < \infty; \ 0 < \sigma < \infty\}. \qquad (3.20)$$

The risk function of the equivariant estimator $E_\rho\{L(\hat{\mu}_\Psi, \mu)\}$ is the same for all $(\mu, \sigma)\epsilon\Theta_\rho$. Let Ψ_ρ be the function that minimizes

$$E_{(0,1,\rho)}\{(\hat{\mu}_\Psi - \mu)^2 | Z_{x,n}, Z_{y,n}\} = E_{(0,1,\rho)}\{\bar{X}_n^2 | Z_{x,n}, Z_{y,n}\}$$
$$+ 2\Psi_\rho E_{(0,1,\rho)}\{\bar{X}_n(\bar{Y}_n - \bar{X}_n) | Z_{x,n}, Z_{y,n}\}$$
$$+ \Psi_\rho^2 E_{(0,1,\rho)}\{(\bar{Y}_n - \bar{X}_n)^2 | Z_{x,n}, Z_{y,n}\}.$$

The minimizer is

$$\Psi_\rho^0(Z_{x,n}, Z_{y,n}) = \frac{-E_{(0,1,\rho)}\{\bar{X}_n(\bar{Y}_n - \bar{X}_n)|Z_{x,n}, Z_{y,n}\}}{E_{(0,1,\rho)}\{(\bar{Y}_n - \bar{X}_n)^2|Z_{x,n}, Z_{y,n}\}}. \tag{3.21}$$

For simplicity of notation we define the variables, $U_1 = \bar{X}_n, U_2 = (\bar{Y}_n - \bar{X}_n)$, $U_3 = Z_{x,n}$, and $U_4 = Z_{y,n}$. The Jacobian of this transformation is U_2^4. Furthermore, the conditional distribution of (U_3, U_4), given U_2, is when $n = 2k + 1$,

$$f_{U_3,U_4}(z_1, z_2|U_2) \propto U_2^{4k} \exp\left(-\frac{1}{2}\left(\frac{z_1 + z_2}{\rho}\right) U_2^2\right). \tag{3.22}$$

Moreover, marginally $U_2 \sim N(0, 1 + \rho)$. Therefore, by Bayes theorem,

$$f_{U_1,U_2}(u_1, u_2|Z_1 = z_1, Z_2 = z_2)$$

$$= \frac{\frac{n}{2\pi\sqrt{\rho}} u_2^{4k} \exp\left(-\frac{1}{2}\left(\frac{z_1+z_2}{\rho}\right) u_2^2 - \frac{n}{2\rho}(u_2 + u_1)^2 - \frac{n}{2} u_1^2\right)}{\frac{\sqrt{n}}{\sqrt{2\pi(1+\rho)}} \int_{-\infty}^{\infty} u_2^{4k} \exp\left(-\frac{1}{2}\left(\frac{z_1+z_2}{\rho}\right) + \frac{n}{1+\rho} u_2^2\right) du_2}. \tag{3.23}$$

Moreover,

$$E_\rho\{U_2^2|Z_1, Z_2\} = \frac{\int_{-\infty}^{\infty} u^{4k+2} \exp\left(-\frac{1}{2}\left(\frac{Z_1+Z_2}{\rho} + \frac{n}{1+\rho} u^2\right) u^2\right) du}{\int_{-\infty}^{\infty} u^{4k} \exp\left(-\frac{1}{2}\left(\frac{Z_1+Z_2}{\rho} + \frac{n}{1+\rho} u^2\right) u^2\right) du}$$

$$= \frac{2n - 1}{\frac{Z_1+Z_2}{\rho} + \frac{n}{1+\rho}}, \tag{3.24}$$

since $E\{N(0, 1)^{2l}\} = (2l)!/(2^l l!)$ for all $l = 0, 1, \ldots$

We see that the locally best location and scale equivariant estimator is a function of ρ, and if ρ is unknown there is no best equivariant estimator. We consider instead Bayes equivariant estimators by adopting a prior distribution $H(\rho)$ for ρ. The marginal density of (Z_1, Z_2) under ρ is

$$f(z_1, z_2; \rho) \propto \frac{1}{\rho\sqrt{1+\rho}} z_1^{k-1} \left(\frac{z_2}{\rho}\right)^{k-1} \left(\frac{z_1 + z_2}{\rho} + \frac{n}{1+\rho}\right)^{-(n-3/2)}. \tag{3.25}$$

Thus, by Bayes theorem, the posterior density of ρ given (Z_1, Z_2) is

$$h(\rho|z_1, z_2) = \frac{h(\rho)f(z_1, z_2; \rho)}{\int_{\rho=0}^{\infty} h(\rho)f(z_1, z_2; \rho)d\rho}. \tag{3.26}$$

The Bayesian equivariant estimator of μ is

$$\hat{\mu}_H = \bar{X}_n + (\bar{Y}_n - \bar{X}_n)\Psi_H \tag{3.27}$$

Table 3.5 Simulation estimates

ρ	n	$E\{\Psi_H\}$	$V\{\Psi_H\}$	$E\{\hat{\mu}_H\}$	$V\{\hat{\mu}_H\}$
2	11	0.39824	0.009350	0.002332	0.066358
2	21	0.37316	0.007350	0.001262	0.031786
2	31	0.36244	0.004944	−0.001102	0.021572
3	11	0.33940	0.007106	0.003939	0.072473
3	21	0.30220	0.004756	0.001407	0.041299
3	31	0.28825	0.003481	−0.000060	0.025209

where,

$$\Psi_H = \frac{\int_{\rho=0}^{\infty} \left((1+\rho)\rho^k \sqrt{1+\rho}\right)^{-1} \left(\frac{Z_1+Z_2}{\rho} + \frac{n}{1+\rho}\right)^{-n+3/2} h(\rho)d\rho}{\int_{\rho=0}^{\infty} \left(\rho^k \sqrt{1+\rho}\right)^{-1} \left(\frac{Z_1+Z_2}{\rho} + \frac{n}{\rho+1}\right)^{-n+3/2} h(\rho)d\rho}. \tag{3.28}$$

In Table 3.5 we present simulation estimates of $E\{\Psi_H(Z_1, Z_2)\}$ and $V\{\Psi_H(Z_1, Z_2)\}$, $E\{\hat{\mu}_H\}$, $V\{\hat{\mu}_H\}$ computed with the prior exponential density $h(\rho) = \frac{1}{2}e^{-\rho/2}$, and 1000 replicas.

We see that the variance of $\hat{\mu}_H$ is close to that of Bayesian estimator $\hat{\mu}_B$ (Table 3.3) but slightly greater than that of T_n (Table 3.1).

3.3 Minimax Estimators

In continuation of the research about estimators of the common mean of two normal distribution, Zacks showed in his (1970a, b, c) paper that the sample mean $\bar{X}_n$ is a minimax estimator, under the invariant loss function $L(\hat{\mu}_n, \mu) = (\hat{\mu}_n - \mu)^2/\sigma_x^2$. This result was improved and generalized in (1974) by Brown and Cohen, who developed several formulas for families of minimax estimators. See also Cohen and Sackrowitz (1974). Several more papers were published later on this subject. We present here the results of Kubokawa (1987a,b) on this subject. He considered the family of location equivariant estimators of the form $\hat{\mu}_{LE} = \bar{X}_n + \phi(\bar{Y}_n - \bar{X}_n, Q_x, Q_y)(\bar{Y}_n - \bar{X}_n)$. Kubokawa proved that the following equivariant estimators

$$\hat{\mu}_M(a, b, c) = \frac{\bar{X}_n + a(\bar{Y}_n - \bar{X}_n)}{(1 + W^*\phi(Q_x, Q_y, (\bar{Y}_n - \bar{X}_n)^2))}, \tag{3.29}$$

where

$$W^* = \frac{(bQ_y + c(\bar{Y}_n - \bar{X}_n)^2)}{Q_x} \tag{3.30}$$

Table 3.6 Simulation
variance of the Brown-Cohen
estimator

ρ	n	$V\{\hat{\mu}_{M_BC}\}$
2	10	0.092945
2	20	0.041738
2	30	0.026131
2	50	0.014901
2	100	0.007356

are minimax, provided the following conditions hold:

1. $0 < a \le 2$, and $b \ge c \ge 0$
2. $n > 2t + 5$ if $c = 0$, or $n > 2t + 2$ if $c > 0$, for some $t \ge 0$
3. The function ϕ / W^{*t} is nondecreasing in Q_x and nonincreasing in Q_y, and when $c > 0$ ϕ / W^{*t} is decreasing in $(\bar{Y}_n - \bar{X}_n)^2$
4. $1/\phi(Q_x, Q_y, (\bar{Y}_n - \bar{X}_n)^2) \le \frac{2n-2t-2}{a(n+2t-1)} u(b, c, ; t)$

where

$$u(b, c; t) = \frac{E\{(b(1 - U) + cU)^{-(t+1)}\}}{E\{b(1 - U) + cU)^{-(t+2)}\}}; \text{and} U \sim \beta\left(3/2, \frac{n-1}{2}\right).$$

Special cases are:

1. $\hat{\mu}_M(0, 0, 0)$ in Zacks (1970a)
2. $\hat{\mu}_M(1, 1, 0)$ in Graybill and Deal (1959)
3. $\hat{\mu}_M(1.39, (n - 1)/(n + 2), 1/(n + 2))$ in Brown and Cohen (1974).

More specifically, the Brown Cohen minimax estimator is of the form

$$\hat{\mu}_{M_BC} = \frac{\bar{X}_n + 1.39(\bar{Y}_n - \bar{X}_n)}{\left(1 + \frac{n-1}{n+2}R + \frac{(\bar{Y}_n - \bar{X}_n)^2}{n+2}\right)}. \tag{3.31}$$

They proved that the variance of this minimax estimator is smaller or equal to that of $V_\rho\{\bar{X}_n\}$ for all ρ, with strict inequality for at least on value of ρ. The variance of this estimator is not necessarily better than that of other estimators as is shown by comparing Table 3.6 with the previous ones in this chapter.

Notice that this table shows that the variance of the Brown-Cohen estimator is larger than that of $\hat{\mu}_H$ (Table 3.5).

3.4 Multivariate Extension

The following multivariate extension was done by Keller and Olkin in (2004). Consider k estimators, $T_1, \ldots, T_k$, of the common mean μ (from k different samples of equal size n), which might be correlated. More specifically, let $\mathbf{T} = (T_1, \ldots, T_k)'$

have a multivariate normal distribution $N(\boldsymbol{\mu}, \boldsymbol{\Sigma})$, where $\boldsymbol{\mu} = \mu 1_k$ and $\boldsymbol{\Sigma}$ is a variance-covariance matrix, whose elements are $\sigma_{ij} = cov(T_i, T_j)$. Moreover, 1_k is a k-dimensional vector of 1's. Let S be a sample covariance matrix having a Wishart distribution $W(\boldsymbol{\Sigma}; k, n)$. Furthermore, $\mathbf{T}$ and S are independent. If $\boldsymbol{\Sigma}$ is known, the best linear unbiased estimator of $\boldsymbol{\mu}$ is

$$\hat{\mu} = \frac{1'_k \boldsymbol{\Sigma}^{-1} \mathbf{T}}{1'_k \boldsymbol{\Sigma}^{-1} 1_k}, \tag{3.32}$$

having a variance $V\{\hat{\mu}\} = 1/1'_k \boldsymbol{\Sigma}^{-1} 1_k$. When $\boldsymbol{\Sigma}$ is unknown, we consider the estimator

$$\hat{\mu}_S = \frac{1'_k S^{-1} \mathbf{T}}{1'_k S^{-1} 1_k}. \tag{3.33}$$

This estimator is unbiased and had the variance

$$V\{\hat{\mu}_S\} = E\left(\frac{1'_k S^{-1} \boldsymbol{\Sigma} S^{-1} 1_k}{1'_k S^{-1} 1_k} \right). \tag{3.34}$$

Keller and Olkin proved that $V\{\hat{\mu}_S\} = \left(\frac{n-1}{n-k}\right) V\{\hat{\mu}\}$.

Cohen (1976) generalized the study to the estimation of common location parameters of two distributions (not necessarily normal) which belong to the same family. In particular he considered the common location parameter of two uniform distributions. George (1991) showed that Stein's type shrinkage estimators dominate the other multivariate estimators and are robust with respect to incorrect values of ρ.

3.5 Estimating the Common Variance of Correlated Normal Distributions

Consider several normal distributions having a common variance. Given independent samples from these distributions, the usual estimator of the common variance is the so-called pooled variance, which is a weighted average of the sample estimators of the variance. This is a common practice in Analysis of Variance. The question is, what estimator should be applied if the samples are not independent. There is no best estimator generally. The results presented in this section are based on the papers of Zacks and Ramig (1987), and Ghezzi and Zacks (2005).

Let $\mathbf{X}_1, \ldots, \mathbf{X}_n$ be n independent (mx1) vectors having identical normal distributions, with zero means, and equicorrelated covariance matrices $V[\mathbf{X}] = \sigma^2 (1 - \rho)I + \rho J$; J is an (mxm) matrix of 1's, and $\frac{-1}{m-1} < \rho < 1$. ρ is the correlation between any two different components of $\mathbf{X}$. Let H be the orthogonal

Helmert matrix, i.e.,

$$
H = \begin{pmatrix}
\frac{1}{\sqrt{m}} & \frac{1}{\sqrt{m}} & \cdots & \cdots & \frac{1}{\sqrt{m}} \\
\frac{1}{\sqrt{2}} & \frac{-1}{\sqrt{2}} & 0 & \cdots & 0 \\
\frac{1}{\sqrt{6}} & \frac{1}{\sqrt{6}} & \frac{-2}{\sqrt{6}} & 0 & 0 \\
\cdots & \cdots & \cdots & \cdots & \cdots \\
\frac{1}{\sqrt{m(m-1)}} & \frac{1}{\sqrt{m(m-1)}} & \cdots & \frac{1}{\sqrt{m(m-1)}} & \frac{-(m-1)}{\sqrt{m(m-1)}}
\end{pmatrix}.
$$

Define $\mathbf{Y}_i = H\mathbf{X}_i$, for $i = 1, \ldots, n$. All the $\mathbf{Y}$ vectors are i.i.d. Normally distributed with zero means, and $V\{Y_1\} = \sigma^2(1 + (m - 1)\rho)$, $V\{(Y_j\} = \sigma^2(1 - \rho)$ for $j = 2, \ldots, m$. Accordingly, the likelihood function of (σ^2, ρ) given $\mathbf{Y}_i, i = 1, \ldots, n$ is

$$
L(\sigma^2, \rho) = \left[\frac{1}{\sigma^{nm/2}(1 - \rho)^{n(m-1)/2}(1 + (m - 1)\rho)^{n/2}} \right]
$$
$$
\times \exp\left[-\frac{1}{2\sigma^2}\left(\frac{V_1}{1 + (m - 1)\rho} + \frac{V_2}{1 - \rho} \right) \right], \tag{3.35}
$$

where

$$
V_1 = \sum_{i=1}^{n} Y_{i1}^2 \text{ and } V_2 = \sum_{i=1}^{n}\sum_{j=2}^{m} Y_{ij}^2.
$$

Notice that

1. V_1 is independent of V_2
2. V_1 is distributed like $\sigma^2[1 + (m - 1)\rho]\chi^2[n]$
3. V_2 is distributed like $\sigma^2(1 - \rho)\chi^2[n(m - 1)]$.

Furthermore, V_1 and V_2 are complete sufficient statistics. The MLEs of σ^2 and ρ are:

$$
\hat{\sigma}_n^2 = \frac{(V_1 + V_2)}{nm} \tag{3.36}
$$

$$
\hat{\rho}_n = (V_1 - V_2/(m - 1))/(V_1 + V_2). \tag{3.37}
$$

These MLEs are strongly consistent and asymptotically normal. For large sample size n their covariance matrix is approximately equal to the inverse of the Fisher information matrix, i.e.,

$$
I^{-1}(\sigma^2, \rho) = \begin{bmatrix} 2\sigma^4(1 + (m - 1)\rho^2) & 2\sigma^2\rho(1 - \rho)(1 + (m - 1)\rho) \\ \cdot & 2(1 - \rho)^2(1 + (m - 1)\rho)^2/(m - 1) \end{bmatrix}/nm. \tag{3.38}
$$

3.5.1 Scale Equivariant Estimators of σ^2

Notice that the estimator (3.36) is scale equivariant. The estimator (3.37) is, on the other hand, invariant. Accordingly, any scale equivariant estimator of σ^2 is of the form

$$\hat{\sigma}^2_{\psi} = T\psi(\hat{\rho}_n). \tag{3.39}$$

The function ψ which minimizes the MSE of $\hat{\sigma}^2_{\psi}$ is

$$\psi^0 = E_{1,\varrho}\{T|\hat{\rho}_n\}/E_{1,\varrho}\{T^2|\hat{\rho}_n\}. \tag{3.40}$$

One can prove that

$$E_{1,\varrho}\{T|\hat{\rho}_n\} = \frac{(nm)^2}{\phi(\hat{\rho}_n, \rho)} \tag{3.41}$$

and

$$E_{1,\varrho}\{T^2|\hat{\rho}_n\} = \frac{(nm)^3(nm+2)}{\phi^2(\hat{\rho}_n, \rho)}. \tag{3.42}$$

Thus

$$\psi^0 = \frac{\phi(\hat{\rho}_n, \rho)}{[nm(nm+2)]}, \tag{3.43}$$

where

$$\phi(\hat{\rho}_n, \rho) = \frac{n(1-\rho)(1+(m-1)\hat{\rho}_n) + (m-1)(1-\hat{\rho}_n)(1+(m-1)\rho)}{(1-\rho)(1+(m-1)\rho)}. \tag{3.44}$$

This function depends on the invariant parameter ρ. Thus, if ρ is unknown, there is no best scale equivariant estimator. The MLE of ϕ is $\phi(\hat{\rho}_n, \hat{\rho}_n) = nm$; substituting this in (3.39) we obtain the estimator

$$\hat{\sigma}^2_{EQ} = \frac{T}{nm+2}. \tag{3.45}$$

The bias of this estimator is

$$E\{\hat{\sigma}^2_{EQ} - \sigma^2\} = \frac{-2\sigma^2}{nm+2} \tag{3.46}$$

and its MSE is

$$E\{(\hat{\sigma}_{EQ}^2 - \sigma^2)^2\} = \frac{2\sigma^4}{(nm+2)^2}(2 + nm(1 + (m-1)\rho^2). \tag{3.47}$$

We show now that the variance of the equivariant estimator (3.45) is uniformly smaller than that of the MLE (3.36), for all ρ.

Indeed,

$$MSE\{\hat{\sigma}_{EQ}^2\}/V\{\hat{\sigma}_n^2\} = \frac{2nm}{(nm+2)^2(1+(m-1)\rho^2)} + \frac{(nm)^2}{(nm+2)^2}$$

$$\leq \frac{(nm)^2 + 2nm}{(nm+2)^2} < 1. \tag{3.48}$$

Another scale equivariant estimator of σ^2 is

$$\hat{\sigma}_{EQ,2}^2 = \frac{T}{nm + 2(1 + (m-1)\hat{\rho}_n^2)}. \tag{3.49}$$

Applying the delta method, we can show that

$$E\{\hat{\sigma}_{EQ,2}^2\} = \frac{nm\sigma^2}{nm + 2(1 + (m-1)\rho^2)} + O\left(\frac{1}{n}\right), \tag{3.50}$$

and

$$MSE\{\hat{\sigma}_{EQ,2}^2\}$$
$$\approx 2\sigma^4 \left[\frac{nm + 2 + nm(m-1)\rho^2 - 4(m-1)(m-2)\rho^3 + 6(m-1)^2\rho^4}{(nm + 2(1 + (m-1)\rho^2)^2}\right]. \tag{3.51}$$

Numerical computations and comparison with simulation results show that the estimator (3.49) is apparently more efficient than (3.45).

3.5.2 Bayesian Equivariant Estimators

We consider here invariant improper-prior measure $d\mu(\sigma^2, \rho) = \frac{1}{\sigma^2}d\sigma^2 d\rho$, and the loss function $\frac{(\hat{\sigma}^2 - \sigma^2)^2}{\sigma^4}$. Let $h(\sigma^2, \rho|t, \hat{\rho})$ denote the posterior density of (σ^2, ρ), i.e.

$$h(\sigma^2, \rho|t, \hat{\rho}) = \frac{f(\sigma^2, \rho)}{\int_{-1/(m-1)}^{1} f(\sigma^2, \rho)d\sigma^2 d\rho},$$

where

$$f(\sigma^2, \rho) = \frac{\exp\left(-\frac{1}{2\sigma^2 nm}\phi(\hat{\rho}, \rho)\right)}{\sigma^{nm+2}(1-\rho)^{n(m-1)/2}(1+(m-1)\rho)^{n/2}}. \tag{3.52}$$

The Bayesian estimator is

$$E\{1/\sigma^2|T, \hat{\rho}\}/E\{1/\sigma^4|T, \hat{\rho}\} = E\left\{\frac{1}{\sigma^2}|T, \hat{\rho}\right\} = E\left\{\frac{1}{\sigma^4}|T, \hat{\rho}\right\} = \frac{T\Psi_B(\hat{\rho})}{(nm+2)}, \tag{3.53}$$

where

$$\Psi_B(\hat{\rho}) = \frac{\int_{\frac{-1}{m-1}}^{1} \phi(\hat{\rho}, \rho)^{\frac{-nm}{2}-1}(1-\rho)^{\frac{-n(m-1)}{2}}(1+(m-1)\rho)^{-n/2}d\rho}{nm\int_{\frac{-1}{m-1}}^{1} \phi(\hat{\rho}, \rho)^{\frac{-nm}{2}-2}(1-\rho)^{\frac{-n(m-1)}{2}}(1+(m-1)\rho)^{-n/2}d\rho}. \tag{3.54}$$

As shown in Ghezzi and Zacks (2005), the equivariant estimator (3.49) has smaller MSE than the Bayesian estimator (3.53), when $\rho = 0.1$. On the other hand, for $\rho = 0.5, 0.9$ the Bayesian estimator has a smaller MSE. There is additional information on equivariant estimators when the covariance matrix is not necessarily equicorrelated.

Chapter 4
Survival Probabilities in Crossing Fields with Absorption Points

As described in Sect. 1.4 of the Introduction, we present here papers dealing with the survival probabilities of objects crossing a field having absorption points at unknown locations. We start with the paper of Zacks and Goldfarb (1966).

4.1 Probabilistic Models

Consider a square A, having an area of size $N[\text{m}^2]$. Absorption points are dispersed at random in this field. k objects (particles) have to cross the field sequentially. If an object encounters an absorption point, it is absorbed with probability p. During an absorption, both the object and the absorption point are destroyed (killed). If the object is not absorbed, the absorption point might absorb another crossing object. We develop in this section the survival probabilities of the crossing objects, or the distribution of the number of objects absorbed (killed), among the k original ones.

We consider two probability models. The first one is that absorption points constitute a Poisson random field. The second model is that the field is Binomial. We remark that these two models are not the only possible ones.

In a Poisson random field, if B_1 and B_2 are any two disjoint sets contained in A, then the number of absorption points in these sets, J_1 and J_2, are two independent random variables, having Poisson distributions, with parameters $\lambda(B_i)$, $i = 1, 2$, where $\lambda(B_i) = (|B_i|/N)\lambda$; $|B|$ is the area of B; λ is the expected number of absorption points in A.

In a Binomial model, the total number of absorption points dispersed in A is known to be $M, M < N$. Let L be a linear path in A, having an area of $|L| = n$. Let J denote the number of absorption points in L. Moreover, $P\{J_L = j\} = \binom{n}{j}(M/N)^j(1 - M/N)^{n-j}$, $j = 0, 1, \ldots, n$. Notice that according to the Poisson field model, $P\{J_L = j\} = e^{-\lambda_L}\lambda_L^j/j!$, $j = 0, 1, 2, \ldots \lambda_L = (n/N)\lambda$.

© Springer Nature Switzerland AG 2020
S. Zacks, *The Career of a Research Statistician*, Statistics for Industry, Technology, and Engineering, https://doi.org/10.1007/978-3-030-39434-9_4

4.2 Survival Probabilities

Let $S^{(i)}, i = 1, 2, \ldots$ denote the event that the i-th object crossing the field in L is survived. Let K_k denote the number of objects absorbed, among the first k crossings. Obviously, $K_0 = 0$, and $K_k \leq k$. Let $f_J^{(0)}(j)$ denote the prior distribution of J, and $f_J^{(k)}(j)$ the posterior distribution of J after k crossing trials. Moreover, $P\{S^{(1)}|J = j\} = (1 - p)^j$. Hence,

$$P\{S^{(1)}\} = \sum_j (1 - p)^j f_J^{(0)}(j) = e^{-\lambda_L p}, \text{ In Poisson Model}$$

$$= (1 - pM/N)^n, \text{ Binomial Model.} \tag{4.1}$$

In the special case of $N = 100$, $M = 20$, $n = 10$, and $p = 0.7$, we get for the Poisson model $P\{S^{(1)}\} = 0.2466$, and for the Binomial model $P\{S^{(1)}\} = 0.2213$.

We develop now the conditional probability $P\{S^{(2)}|S^{(1)}\}$.

By Bayes theorem, the prior density $f_J^{(0)}(j)$ becomes, given $S^{(1)}$, the posterior density

$$f_J^{(1)}(j|K_1 = 0) = \frac{I\{0 \leq j \leq n\}(1 - p)^j f_J^{(0)}(j)}{\sum_{j=0}^n (1 - p)^j f_J^{(0)}(j)}. \tag{4.2}$$

Accordingly, $P\{S^{(2)}|S^{(1)}\} = \sum_{j=0}^n (1 - p)^j f_J^{(1)}(j|K_1 = 0)$.

Thus, according to the Binomial Model we get

$$P\{S^{(2)}|S^{(1)}\} = \frac{(1 - p(2 - p)M/N)^n}{(1 - pM/N)^n}. \tag{4.3}$$

For example, if $N = 100$, $M = 20$, $n = 10$, $p = 0.7$, $P\{S^{(2)}|S^{(1)}\} = 0.6061$. Indeed, if the first object crosses the field successfully, the probability of survival of the second object increases considerably.

We develop now the conditional probability of survival of the second object, given that the first one was absorbed, i.e., $P\{S^{(2)}|K_1 = 1\}$. Since $K_1 = 1$, the number of absorption points was at least 1. Second, since the trial stops immediately after an absorption, the number of successful crossings until an absorption has a geometric distribution. Thus, for the Binomial model, we follow the following steps: We start with the posterior density of J, given that $K_1 = 1$, which is

$$f_J^{(1)}(j|K_1 = 1) = \frac{I\{1 \leq j \leq n\}(1 - (1 - p)^j)f_J^{(0)}(j + 1)}{\sum_{j=0}^{n-1}(1 - p)^j f_J^{(0)}(j)}. \tag{4.4}$$

Accordingly,

$$P\{S^{(2)}|K_1 = 1\} = \frac{\sum_{j=0}^{n-1}(1-p)^j(1-(1-p)^{j+1})f_J^{(0)}(j+1)}{\sum_{j=0}^{n-1}(1-(1-p)^{j+1})f_J^{(0)}(j+1)}. \tag{4.5}$$

Hence, for the Binomial model we get

$$P\{S^{(2)}|K_1 = 1\} = \frac{1}{1-p}\Big[(1-pM/N)^n$$

$$-(1-p(2-p)M/N)^n\Big]/\Big[1-(1-pM/N)^n\Big]. \tag{4.6}$$

When $N = 100, M = 20, n = 10, p = 0.7$, we obtain $P\{S^{(2)}|K_1 = 1\} = 0.37314$. In addition we also get that $P\{S^{(2)}\} = 0.4247$.

In a similar fashion we can prove that in the Binomial Model

$$P\{S^{k+1}|K_k = 0\} = \frac{(1-p\sum_{l=1}^{k+1}(-1)^{l-1}\binom{k+1}{l})p^{l-1}M/N)^n}{(1-p\sum_{l=1}^{k}(-1)^{l-1}\binom{k}{l})p^{l-1}M/N)^n}. \tag{4.7}$$

The case of $K_k > 1$ is more complicated. For example,

$$f_J^{(2)}(j|K_2 = 2)$$

$$= \frac{(1-(1-p)^j)(1-(1-p)^{j-1})f_J^{(0)}(j)I\{j \geq 2\}}{1-\sum_{j=2}^{\infty}(2-p)(1-p)^{j-1}f_J^{(0)}(j)+\sum_{j=2}^{\infty}(1-p)^{2j-1}f_J^{(0)}(j)}. \tag{4.8}$$

Hence,

$$P\{S^{(3)}|K_2 = 2\}$$

$$= \frac{\sum_{j=2}^{\infty}(1-p)^{j-2}f_J^{(0)}(j)-\sum_{j=2}^{\infty}(2-p)(1-p)^{2j-3}f_J^{(0)}(j)+\sum_{j=2}^{\infty}(1-p)^{3j-3}f_J^{(0)}(j)}{1-\sum_{j=2}^{\infty}(2-p)(1-p)^{j-1}f_J^{(0)}(j)+\sum_{j=2}^{\infty}(1-p)^{2j-1}f_J^{(0)}(j)}. \tag{4.9}$$

The Binomial model yields, for $N = 100, M = 20, n = 10, p = 0.7$, the result $P\{S^{(3)}|K_2 = 2\} = 0.29429$.

For $K_4 = 4$ we obtain

$$f_J^{(4)}(j|K_2 = 4) = \frac{\prod_{l=0}^{3}(1-(1-p)^{j-l})]f_J^{(0)}(j)I\{j \geq 4\}}{\sum_{j=4}^{n}[\prod_{l=0}^{3}(1-(1-p)^{j-l})]f_J^{(0)}(j)}. \tag{4.10}$$

Notice that in this case,

$$\prod_{l=0}^{3}(1-(1-p)^{j-l})$$

$$= 1-(1-p)^{j}\sum_{l=0}^{3}(1/(1-p)^{l})+(1-p)^{2j}(1/(1-p)^{3}+\sum_{l=1}^{5}(1/(1-p)^{l}))$$

$$-(1-p)^{3j}\sum_{l=3}^{6}(1/(1-p)^{l})+(1-p)^{4j-6}. \qquad (4.11)$$

Finally one can get $P\{S^{(5)}|K_4 = 4\} = \sum_{j=4}^{n}(1-p)^{j-4}f_J^{(4)}(j|K_4 = 4)$. For the Binomial Model, if $N = 100, M = 20, n = 10, p = 0.7$, we get $P\{S^{(5)}|K_4 = 4\} = 0.738070$, and if $M = 30$ we get $P\{S^{(5)}|K_4 = 4\} = 0.592619$.

For the Poisson field model, with $\lambda = 2.154613$, we obtain the following numerical results:

$$P\{S^{(1)}\} = 0.22130$$

$$P\{S^{(2)}|S^{(1)}\} = 0.636056$$

$$P\{S^{(2)}|K_1 = 1\} = 0.315686$$

$$P\{S^{(2)}\} = 0.636056 * 0.22130 + 0.315686 * (1-0.22130) = 0.386103$$

$$P\{S^{(5)}|K_4 = 4\} = 0.616475.$$

When $1 \leq K_k < k$, the survival probabilities depend on the order of events. If $D^{(i)}$ denotes that the object in the i-th crossing was destroyed, the posterior densities of J, given $\{S^{(1)}, D^{(2)}\}$ or $(D^{(1)}, S^{(2)})$, are different. Indeed, in these cases,

$$P\{S^{(3)}|S^{(1)}, D^{(2)}\} = \frac{\sum_{j=1}^{\infty}(1-p)^{2j-1}(1-(1-p)^{j})f_J^{(0)}(j)}{\sum_{j=1}^{\infty}(1-p)^{j}(1-(1-p)^{j})f_J^{(0)}(j)}, \qquad (4.12)$$

and

$$P\{S^{(3)}|D^{(1)}, S^{(2)}\} = \frac{\sum_{j=1}^{\infty}(1-p)^{2j-2}(1-(1-p)^{j})f_J^{(0)}(j)}{\sum_{j=1}^{\infty}(1-p)^{j-1}(1-(1-p)^{j})f_J^{(0)}(j)}.$$

In the paper of Zacks (1967), the survival probabilities are given in terms of the generating functions of J. These are good approximations when J has a Poisson distribution. The generating function for the Poisson distribution, with parameter λ, is $G(s) = e^{-\lambda}\sum_{j=0}^{\infty}s^j\lambda^j/j! = \exp(-\lambda(1-s))$, for $0 < s < 1$. Let $s =$

$1 - p$. Then, the survival function $E\{(1 - p)^J\} = G(s)$. Let $K_k = d$, then we can approximate formulas (4.12) and (4.13), when the prior distribution is Poisson, with

$$P\{S^{(4)}|K_2 = 1\} = \frac{(G(s^3) - G(s^4))}{[s(G(s^2) - G(s^3))]}. \tag{4.13}$$

In the original (4.12), Poisson distribution with $\lambda = 2.154613$ yields $P\{S^{(3)}|S^{(1)}, D^{(2)}\} = 0.703509$. Formula (4.13) yields $P\{S^{(3)}|D^{(1)}, S^{(2)}\} = 0.6368$. The approximation (4.14) gives $P\{S^{(3)}|K_3 = 1\} = 0.696604$. The following approximations are given in Zacks (1967)

$$P\{S^{(4)}|K_3 = 0\} = \frac{G(s^4)}{G(s^3)},$$

$$P\{S^{(4)}|K_3 = 1\} = \frac{(G(s^3) - G(s^4))}{[s(G(s^2) - G(s^3))]},$$

$$P\{S^{(4)}|K_3 = 2\} = \frac{[sG(s^2) - (1+s)G(s^3) + G(s^4)]}{[s^2(sG(s) - (1+s)G(s^2) + G(s^3))]}$$

$$P\{S^{(4)}|K_3 = 3\} = \frac{[s^3(1-s)G(s) - s(1-s^3)G(s^2) + (1-s^3)G(s^3) - (1-s)G(s^4)]}{[s^3(s^3(1-s) - s(1-s^3)G(s) + (1-s^3)G(s^2) - (1-s)G(s^3))]}.$$

4.3 Bayesian Strategies for Choosing Alternative Paths

Suppose that the field has two alternative crossing paths. There are n objects that should cross the field sequentially. The problem is how to direct the objects on which path to cross, when the number of absorption points in the two paths J_1 and J_2 are unknown. Let $M_k = m$ denote the number of objects, among the first k crossings that were directed to path 1. $0 \leq m \leq k$. Let $K_k^{(1)} = d_1$ and $K_k^{(2)} = d_2$ denote the number of objects killed on path 1 and path 2, respectively, among the first k crossings. We consider now a joint prior density $\xi_0(j_1, j_2)$ for (J_1, J_2). Notice that $d_1 \leq m, d_2 \leq k - m$. The posterior density is

$$\xi_k(j_1, j_2|m, d_1, d_2)$$

$$= \frac{\prod_{l_1=0}^{d_1-1}(1 - s^{j_1-l_1}) \prod_{l_2=0}^{d_2-1}(1 - s^{j_2-l_2})s^{(m-d_1)(j_1-d_1)+(k-m-d_2)(j_2-d_2)}}{E\left\{\prod_{l_1=0}^{d_1-1}(1 - s^{j_1-l_1}) \prod_{l_2=0}^{d_2-1}(1 - s^{j_2-l_2})s^{(m-d_1)(j_1-d_1)+(k-m-d_2)(j_2-d_2)} \times \xi_0(j_1, j_2)\right\}}.$$

$$\tag{4.14}$$

Notice that $\prod_{l=0}^{-1} a^l = 1$. Also, if J_1 and J_2 are independent, then the joint density is a product of the two marginal densities. That is, $\xi_k(j_1, j_2|m, d_1, d_2) = \xi_m^{(1)}(j_1|d_1)\xi_{k-m}^{(2)}(j_2|d_2)$. Furthermore, under independence, the information of what happens in path 2 is irrelevant for the survival probability in path 1, and vice versa. The optimal strategy is the one that maximizes the expected number of survivors. The problem is similar to the famous "two-armed bandit problem." Since n is finite, one can obtain the optimal strategy according to dynamic programming, by backward induction.

Let $\varsigma_{k+1}^{(i)}(M_k, D_k^{(1)}, D_k^{(2)})$ denote the posterior survival probability of the $(k + 1)$th object , given $(M_k, D_k^{(1)}, D_k^{(2)})$, if crossing in the i-th path, i.e.,
$\varsigma_{k+1}^{(i)}(M_k, D_k^{(1)}, D_k^{(2)}) = E\{s^{J_i - D_k^{(i)}}|(M_k, D_k^{(1)}, D_k^{(2)})\}$.

According to the dynamic programming, the n-th object is directed to the path for which the survival probability is maximal. We define accordingly the function

$$S_n(m, d_1, d_2) = \max_{i=1,2}\{\varsigma_n^{(i)}(m, d_1, d_2)\}. \tag{4.15}$$

Let $\chi_n^{(i)} = 1$ if the n-th object is absorbed in the i-th path ($i = 1, 2$), and $\chi_n^{(i)} = 0$, otherwise. The conditional probability function of $\chi_n^{(i)}$, given $\{M_{n-1} = m, D_{n-1}^{(1)} = d_1, D_{n-1}^{(2)} = d\}$, is

$$q_n^{(i)}(x|m, d_1, d_2) = (\varsigma_n^{(i)}(m, d_1, d_2))^{1-x}(1 - \varsigma_n^{(i)}(m, d_1, d_2))^x. \tag{4.16}$$

We define then

$$S_{n-1}(m, d_1, d_2) = \max_{i=1,2}\{S_{n-1}^{(i)}(m, d_1, d_2)\}, \tag{4.17}$$

where

$$S_{n-1}^{(1)}(m, d_1, d_2) = \varsigma_n^{(1)}(m, d_1, d_2) + \sum_{x=0}^{1} S_n(m + 1, d_1 + x, d_2)q_n^{(1)}(x|m, d_1, d_2) \tag{4.18}$$

and

$$S_{n-1}^{(2)}(m, d_1, d_2) = \varsigma_n^{(2)}(m, d_1, d_2) + \sum_{x=0}^{1} S_n(m, d_1, d_2 + x)q_n^{(2)}(x|m, d_1, d_2). \tag{4.19}$$

If $S_{n-1}^{(1)}(m, d_1, d_2) \geq S_{n-1}^{(2)}(m, d_1, d_2)$ the $(n - 1)$-st object is directed to path 1, and to path 2 otherwise.

In a similar fashion all the other $S_k(m, d_1, d_2)$ functions are constructed , for $k = n-2, n-3, \ldots 1, 0$. This is a tedious process. This decision process is simplified if J_1 and J_2 are priorly independent. Generally, under independence, the myopic

Table 4.1 Posterior survival probabilities and the Bayes survival functions

k	m	d_1	d_2	$\varsigma_{k+1}^{(1)}$	$\varsigma_{k+1}^{(2)}$	$S_{k+1}^{(1)}$	$S_{k+1}^{(2)}$	S_{k+1}
3	3	0	0	0.8036	0.7408	0.8306	0.7408	0.8306
3	3	1	0	0.7936	0.7408	0.7936	0.7408	0.7936
3	3	2	1	0.7822	0.7408	0.7822	0.7408	0.7822
3	3	3	0	0.7740	0.7408	0.7740	0.7408	0.7740
3	2	0	0	0.7843	0.7634	0.7843	0.7634	0.7843
3	2	1	0	0.7733	0.7634	0.7733	0.7634	0.7733
3	2	2	0	0.7618	0.7634	0.7618	0.7634	0.7634
3	2	0	1	0.7843	0.7615	0.7843	0.7515	0.7843
3	2	1	1	0.7618	0.7515	0.7733	0.7515	0.7733
3	2	2	1	0.7618	0.7515	0.7618	0.7515	0.7618
2	2	0	0	0.7843	0.7408	1.5857	1.5251	1.5857
2	2	1	0	0.7733	0.7408	1.5643	1.5141	1.5643
2	2	2	0	0.7618	0.7408	1.5420	1.5038	1.5420
2	1	0	0	0.7634	0.7634	1.5451	1.5451	1.5451
2	1	1	0	0.7515	0.7634	1.5223	1.5451	1.5451
2	1	0	1	0.7634	0.7515	1.5451	1.5223	1.5451
2	1	1	1	0.7515	0.7515	1.5219	1.5219	1.5219
1	1	0	0	0.7634	0.7408	2.3395	2.2859	2.3395
1	1	1	0	0.7515	0.7408	2.3102	2.2799	2.3102
0	0	0	0	0.7408	0.7408	3.0727	3.0727	3.0727

strategy is reasonable if not optimal. In other words, if $\varsigma_{m+1}(d_1) \geq \varsigma_{n+1-m}(d_2)$, then choose path 1, and choose path 2 otherwise.

For example, suppose J_1 and J_2 are independent, having the same Poisson distribution, with parameter $\lambda = 3$, $s = 0.9$, and $n = 4$.

Notice that in this case, $\varsigma_{k+1}^{(1)}(m, d_1, d_2) = \varsigma_{m+1}(d_1) = \varsigma_{k+1}^{(2)}(k - m, d_1, d_2)$. In Table 4.1 we illustrate the dynamic programming computations for this case.

According to Table 4.1, the expected number of survivors, following the optimal strategy, is 3.0727.

4.4 Telegraph Process with Elastic Boundary at the Origin

We discuss in the present section the time till absorption at the origin of a particle which moves up and down intermittently on the positive part of the real line. Whenever the particle hits the origin it is absorbed with probability α, and the process stops or reflected upwards with probability $1 - \alpha$. The motion up is at velocity $V_1 = c$, for a random length of time U, and the motion down is at velocity $V_2 = -d$, for a random time D, repeatedly. Such a process is called a *telegraph process*. Here the process is restricted to the positive part of the line, with velocities $V_1 = 1$ and $V_2 = -1$. The random times up and down constitute an alternating renewal process $\{U_1, D_1, U_2, D_2, \ldots\}$, where all random variables are mutually independent. All the U variables have the same distribution F, and all the D variables have the same distribution G. In Chap. 11 we will further discuss telegraph processes. Applications of these processes are in applied physics, financial mathematics, and more.

Let C denote the time between consecutive visit to the origin. This is the random length of a cycle. Let M denote the random number of cycles until absorption. Clearly M has a geometric distribution with

$$P\{M = m\} = \alpha(1 - \alpha)^{m-1}, m = 1, 2, \ldots$$

If the particle is not absorbed the motions continue with the same stochastic rules as before. The random time till absorption is $A = \sum_{i=1}^{M} C_i$.

Define the compound renewal process

$$Y(t) = \sum_{i=1}^{N(t)} D_i, \tag{4.20}$$

where $N(t) = \max\{n \geq 0 : \sum_{i=1}^{n} U_i \leq t\}$. Clearly, $N(t) = 0$ if $U_1 > t$. The c.d.f. of $Y(t)$ is

$$H(y, t) = \sum_{n=0}^{\infty} (F^{(n)}(t) - F^{(n+1)}(t))G^{(n)}(y). \tag{4.21}$$

Define a stopping time

$$T = \inf\{t > 0 : Y(t) \geq t\}. \tag{4.22}$$

It follows that $C = 2T$. As proved by Stadje and Zacks (2004), the density of T is

$$\varphi(t) = \int_0^t (1 - x/t)h(x, t)(1 - G(t - x))dx, \tag{4.23}$$

where $h(x, t)$ is the density of $H(x, t)$. Accordingly, the density of C is $\varphi_C(t) = (1/2)\varphi(t/2)$. It follows that the density of A is

$$f_A(t) = \alpha \sum_{n=1}^{\infty} (1 - \alpha)^{n-1} \varphi_C^{(n)}(t). \tag{4.24}$$

Table 4.2 Analytical results, $\alpha = 0.1$

λ	μ	$E\{C\}$	$V\{C\}$	$E\{A\}$	$V\{A\}$
2	1.8	10	1900	100	28000
2	1.6	5	225	50	4500
2	1.4	3.33	62.96	33.33	1629.63
2	1.2	2.5	25	25	812.6
2	1	2	12	20	480

Analytic expressions for this density and associated functionals are derived by Di Crescenzo, Martinucci, and Zacks (2018) for the special case where the distributions F and G are exponentials, with parameters λ and μ, respectively. If $\lambda > \mu$, we get

$$E\{C\} = \frac{2}{\lambda - \mu}, \tag{4.25}$$

and

$$V\{C\} = \frac{4(\lambda + \mu)}{(\lambda - \mu)^3}. \tag{4.26}$$

It follows that in this special case,

$$E\{A\} = \frac{2}{\alpha(\lambda - \mu)} \tag{4.27}$$

and

$$V\{A\} = \frac{4(\lambda + \mu(2\alpha - 1))}{\alpha^2(\lambda - \mu)^3}. \tag{4.28}$$

Some numerical results are given in Table 4.2.
For additional information on telegraph processes see Chap. 11.

Chapter 5
Logistics and Operations Analysis for the Military

As described in Sect. 1.5 of the Introduction, we present in this chapter theoretical contributions made in the areas of inventory control, readiness evaluation of big military units, and reliability of items entering a wear out phase. The inventory control study served the needs of naval applications in the 60s and the 70s. Inventory systems today are based on computer control and on fast delivery. What is described here might not be applicable in modern times, but might be of historical interest. We start with one-echelon inventory system and then proceed to two-echelon systems. Before we start with the inventory control theory, we introduce the scenario. The stock of the customer (a ship or a submarine) consists of a large number of items. Each item requires its specific demand analysis. The theory presented in this chapter discusses a special case of discrete variables, which present the number of items consumed between replenishments. The theory for continuous demand variables (fluids) can be modified appropriately.

5.1 One-Echelon Inventory System

The material in this section is based on the paper of Zacks (1969a).

5.1.1 The Inventory Model

The one-echelon system considered represents the type of model appropriate for operational ships (submarines). The important feature in this model is the possible return of excess stock to the supporting base (the tender ship) after each patrol. This assumption combined with a simple cost function yields a relatively simple solution of the Bayes sequential decision problem. In the present section we present

© Springer Nature Switzerland AG 2020
S. Zacks, *The Career of a Research Statistician*, Statistics for Industry, Technology, and Engineering, https://doi.org/10.1007/978-3-030-39434-9_5

the Bayesian analysis when the demand distribution is Poisson, with Gamma prior distribution for the Poisson mean λ. More specifically, the inventory model is:

1. The demand random variables are i.i.d, having a Poisson distribution with mean λ.
2. The prior distribution of λ is Gamma, $G(\upsilon, \beta)$, where β is the scale parameter and υ is the shape parameter.
3. There is no lead time, and orders are replenished instantaneously at the beginning of each period (month).
4. There are no backlogs, i.e., shortages or overstocking can be adjusted at the beginning of each month.

We assume the common convex piece-wise linear cost function

$$L(x; k) = I\{x \leq k\}c(k - x) + I\{x > k\}p(x - k), \tag{5.1}$$

where x is the demand level in a period and k is the stock level at the beginning of a period. $c[\$]$ is the cost of carrying one unit of stock a whole period; $p[\$]$ is the penalty cost for a shortage of one unit. As will be shown, the adjusted stock level at the beginning of a period is the $\frac{p}{c+p}$-th quantile of the posterior distribution of λ, given the demand levels at the previous months. For a planning horizon of N months, the objective is to determine the levels $\{k_1, k_2, \ldots, k_N\}$ that minimize the total expected loss.

5.1.2 Prior, Posterior, and Marginal Distributions

As mentioned above, the prior distribution of λ is the Gamma, $G(\upsilon, \beta)$ with density

$$g(x; \upsilon, \beta) = \frac{1}{\beta^\upsilon \Gamma(\upsilon)} x^{\upsilon-1} e^{-x/\beta}. \tag{5.2}$$

If $\{X_1, \ldots, X_n\}$ are i.i.d. Poisson λ, then the posterior distribution of λ, given $S_n = \sum_{j=1}^{n} X_j$, is the Gamma distribution $G(S_n + \upsilon, \beta/(1 + \beta n))$. Notice that $S_0 = 0$. The predictive distribution of X_{n+1}, given S_n, is the Negative-Binomial distribution, N.B., with density

$$g(x; \varphi_{n+1}, \upsilon_{n+1}) = \frac{\Gamma(x + \upsilon_{n+1})}{\Gamma(x + 1)\Gamma(\upsilon_{n+1})} \varphi_{n+1}^x (1 - \varphi_{n+1})^{\upsilon_{n+1}}, \tag{5.3}$$

where

$$\varphi_{n+1} = \frac{\beta}{(1 + (n + 1)\beta)}; \upsilon_{n+1} = \upsilon + S_n. \tag{5.4}$$

5.1.3 Bayesian Determination of Stock Levels

If $N = 1$, the prior risk function, with the loss function (5.1), is

$$R_1(k; \varphi_1, \upsilon_1) = \sum_{x=0}^{\infty} L(k; x) g(x; \varphi_1, \upsilon_1). \tag{5.5}$$

The optimal stock level k^0 is approximately the $p/(c + p)$ quantile of the N.B. distribution (5.3). Indeed, if the distribution of X, $F(x)$ is continuous, then

$$\frac{\partial}{\partial k} R(k, F) = \frac{\partial}{\partial k} \left[c \int_{x=0}^{k} (k - x) dF(x) + p \int_{x=k}^{\infty} (x - k) dF(x) \right]$$
$$= cF(k) - p(1 - F(x)). \tag{5.6}$$

Equating this partial derivative to zero we get $F(k^0) = \frac{p}{c+p}$. If F is a discrete distribution, the minimizer is approximately the above quantile. Returning to (5.5), if $\beta = 1$, and $\upsilon = 10$, we obtain for $n = 1$ that $\varphi_1 = \frac{1}{3}$ and $\upsilon_1 = 10$. Moreover, if $c = 3$ and $p = 7$, the 0.7 quantile of the corresponding N.B. distribution is $k^0 = 6$.

Let $\varrho_0(\varphi, \upsilon)$ denote the minimal Bayes risk associated with $k^0(\varphi, \upsilon)$. As proved in Appendix 2 of Zacks (1969a), the minimal risk is approximately

$$\varrho_0(\varphi, \upsilon) = (c + p) \left[\frac{\Gamma(k^0(\varphi, \upsilon) + \upsilon + 1)}{\Gamma(\upsilon)} \Gamma(k^0(\varphi, \upsilon) + 1) \right] \varphi^{k^0(\varphi, \upsilon)+1} (1 - \varphi)^{\upsilon - 1}. \tag{5.7}$$

This risk for the above example is $\varrho_0(\frac{1}{3}, 10) = 9.5012$.

We proceed now to the general case of N. The method is the backward induction of Dynamic Programming, which was discussed in Chap. 4. Since there are only N period, whatever was the inventory policy before, the optimal stock level for the last period is $k_N^0(\varphi_N, \upsilon_N)$. We consider now the determination of the optimal stock level for period $N - 1$. We denote it by $k_{N-1}^0(\varphi_{N-1}, \upsilon_{N-1})$. This is the value of k minimizing

$$R_1(k; \varphi_{N-1}, \upsilon_{N-1}) = E\{L(X; k) + \varrho_0(\varphi_N, \upsilon_{N-1} + X) | \varphi_{N-1}, \upsilon_{N-1}\}, \tag{5.8}$$

where X has the Negative-Binomial distribution, $N.B.(\varphi_{N-1}, \upsilon_{N-1})$. We define the Bayes risk for the last 2 months as

$$\varrho_1(\varphi_{N-1}, \upsilon_{N-1}) = \inf_{0 < k < \infty} \{R_1(k; \varphi_{N-1}, \upsilon_{N-1})\}. \tag{5.9}$$

Since the density (5.3) does not depend on the stock level k_{N-1}, the optimal value of k for period $N - 1$ is the $p/(c + p)$ quantile of the N.B. distribution with parameters $\varphi_{N-1}, \upsilon_{N-1}$. Furthermore, the minimal Bayes risk is

$$\varrho_1(\varphi_{N-1}, \upsilon_{N-1}) = \varrho_0(\varphi_{N-1}, \upsilon_{N-1}) + E\{\varrho_0(\varphi_N, \upsilon_{N-1} + X)|\varphi_{N-1}, \upsilon_{N-1}\}. \tag{5.10}$$

For the $N - 2$ period we define

$$R_2(k; \varphi_{N-2}, \upsilon_{N-2}) = E\{L(X; k) + \varrho_1(\varphi_{N-1}, \upsilon_{N-2} + X)|\varphi_{N-2}, \upsilon_{N-2}\}. \tag{5.11}$$

Here again, $k_{N-2}^0(\varphi_{N-2}, \upsilon_{N-2})$ is the $p/(c+p)$ quantile of the N.B. distribution, with parameters $\varphi_{N-2}, \upsilon_{N-2}$. The corresponding minimal risk is

$$\varrho_2(\varphi_{N-2}, \upsilon_{N-2}) = \varrho_0(\varphi_{N-2}, \upsilon_{N-2}) + E\{\varrho_1(\varphi_{N-1}, \upsilon_{N-2} + X)|\varphi_{N-2}, \upsilon_{N-2}\}. \tag{5.12}$$

Generally, for $n = 1, 2, \ldots, N - 1$,

$$k_{N-n}^0(\varphi_{N-n}, \upsilon_{N-n}) = \min\{k : N.B.(k; \varphi_{N-n}, \upsilon_{N-n}) \geq p/(c + p)\}.$$

Furthermore,

$$\varrho_n(\varphi_{N-n}, \upsilon_{N-n}) = \varrho_0(\varphi_{N-n}, \upsilon_{N-n})$$
$$+ E\{\varrho_{n-1}(\varphi_{N-n+1}, \upsilon_{N-n} + X_{N-n+1})|\varphi_{N-n}, \upsilon_{N-n}\}. \tag{5.13}$$

By substitutions we obtain

$$\varrho_n(\varphi_{N-n}, \upsilon_{N-n}) = \varrho_0(\varphi_{N-n}, \upsilon_{N-n})$$
$$+ \sum_{j=1}^{n} E\{\varrho_0(\varphi_{N-n+j}, \upsilon_{N-n} + \sum_{l=1}^{j} X_{N-n+l})|\varphi_{N-n}, \upsilon_{N-n}\}. \tag{5.14}$$

Finally, since the distribution of $S_j^* = \sum_{l=1}^{j} X_{N-n+l}$ is that of the sum of j i.i.d. $N.B.(\varphi_{N-n}, \upsilon_{N-n})$, which is $N.B.(\varphi_{N-n}, j\upsilon_{N-n})$, we get

$$\varrho_n(\varphi_{N-n}, \upsilon_{N-n}) = \varrho_0(\varphi_{N-n}, \upsilon_{N-n})$$
$$+ \sum_{j=1}^{n} E\{\varrho_0(\varphi_{N-n+j}, \upsilon_{N-n} + X)|\varphi_{N-n}, j\upsilon_{N-n}\}. \tag{5.15}$$

In particular, the minimal total Bayes risk is

$$\varrho_N(\varphi, \upsilon) = \varrho_0(\varphi, \upsilon) + \sum_{j=1}^{N} E\{\varrho_0(\varphi_j, \upsilon + X)|\varphi, j\upsilon\}. \tag{5.16}$$

β	υ	$\varrho_{10}(\varphi, \upsilon)$
1	10	55.5796
1	15	61.7121
1	20	80.7777
2	10	70.0978
2	15	86.4489
2	20	92.1944

Table 5.1 Simulated total minimal Bayes risk

In Table 5.1 we present the simulated total minimal Bayes risk for the case of $p = 7, c = 3, N = 10$. The simulation is needed to approximate $E\{\varrho_0(\varphi_j, \upsilon + X)|\varphi, j\upsilon\}$. The computation is done by program "Tbayriskex" in the Appendix.

5.2 Two-Echelon Inventory System

The material in this section is based on the papers Zacks and Fennel (1972), Zacks and Fennel (1974), and Zacks (1974).

5.2.1 The Inventory Model

In the two-echelon system, the higher echelon is the depot, which obtains parts from the manufacturers. The lower echelon consists of several stations (tender ships). The operational boats (submarines) assume the role of customers, which demand at certain random times a random number of units of specified replacement parts.

Let D denote the depot, and let $E_1, \ldots, E_k$ denote k stations. Without loss of generality, we consider a time period of 1 month. Let $X_n^{(i)}, (i = 1, \ldots, k; n = 1, 2, \ldots)$ denote the number of units demanded at station i in period n. We further assume that $X_n^{(i)}$ are independent random variables having Poisson distributions with means $\lambda_i (i = 1, \ldots, k)$. The parameters λ_i are unknown. The random variables $T_n^{(i)} = \sum_{j=1}^n X_j^{(i)}$ are independent having Poisson distributions with means $n\lambda_i (i = 1, \ldots, k)$. These statistics are also minimally sufficient for their respective parameters. In this model we assume that stations in Echelon II can send excessive stock back to the depot. There is no flow of material between stations, and the customers cannot send material back to the stations. We assume the following inventory flow:

1. At the beginning of the n-th month the depot D orders from the manufacturers Z_n units, $Z_n \geq 0$. The lead time is L months ($L \geq 1$). Units ordered from the manufacturer cannot be returned.
2. At the beginning of the n-th month, station E_i of Echelon II orders $Y_n^{(i)}$ units from the depot. The orders arrive at the end of the month. Excess of stock at the

stations can be adjusted by sending units back to the depot at the beginning of a month. In this case the corresponding $Y_n^{(i)} < 0$.
3. There is no flow back of units from the customers (submarines) to the stations. Demand of customers is according to the exact number of failures.

Let Q_n denote the stock level at the depot at the end of the n-th month. Let $S_n^{(i)}$ denote the stock level at station E_i at the end of the n-th month. We consider the following restriction

$$- S_{n-1}^{(i)} \leq Y_n^{(i)} \text{ for all } i = 1, \ldots, k; n = 1, 2, \ldots$$

$$\sum_{i=1}^{k} Y_n^{(i)} \leq Q_{n-1}, \text{ for all } n = 1, 2, \ldots. \tag{5.17}$$

Useful relationships are

$$S_n^{(i)} = (S_{n-1}^{(i)} + Y_n^{(i)} - X_n^{(i)})^+, i = 1, \ldots, k; n = 1, 2, \ldots, N \tag{5.18}$$

and

$$Q_n = \left[Q_{n-1} + Z_{n-L+1} - \sum_{i=1}^{k} Y_n^{(i)} - \sum_{i=1}^{k} (S_{n-1}^{(i)} - X_n^{(i)})^- \right]^+, n = 1, 2, \ldots, N. \tag{5.19}$$

5.2.2   The Cost Structure

Let c_i and p_i be the monthly carrying cost or penalty for shortage of one unit in station E_i, $i = 1, \ldots, k$. Let C_d be the monthly carrying cost of a unit in the depot D, and let P_d be the penalty for shortage of one unit in the depot. For simplicity we assume that all costs are charged at the end of each month. The total inventory cost of the system at the end of the n-th month is

$$L_n (Q_{n-1}, S_{n-1}, Z_{n-L+1}; X_n)$$

$$= \sum_{i=1}^{k} \left[c_i \left(S_{n-1}^{(i)} - X_n^{(i)} \right)^+ + p_i \left(S_{n-1}^{(i)} - X_n^{(i)} \right)^- \right]$$

$$+ C_d \left[Q_{n+1} + Z_{n-L+1} - \sum_{i=1}^{k} Y_n^{(i)} - \sum_{i=1}^{k} \left(S_{n-1}^{(i)} - X_n^{(i)} \right)^- \right]^+$$

$$+ P_d \left[Q_{n+1} + Z_{n-L+1} - \sum_{i=1}^{k} Y_n^{(i)} - \sum_{i=1}^{k} \left(S_{n-1}^{(i)} - X_n^{(i)} \right)^- \right]^-, \quad n = 1, 2, \ldots, N.$$

$$\tag{5.20}$$

The objective is to determine the order levels Y_n and Z_n at the beginning of each month, so as to minimize the expected total cost

$$R = E\left[\sum_{n=1}^{N} L_n\left(Q_{n-1}, S_{n-1}, Z_{n-L+1}; X_n\right)\right]. \tag{5.21}$$

The cost function (5.21) is considerably more complicated than the one (5.1), which is the one-echelon loss. The dynamic programming solution is unpractical. One could consider sub-optimal solution based on the above one-echelon solution. A different simplification was done in the paper of Zacks and Fennel (1973), in which the ordering level of the depot assures that the predictive probability of shortage at the depot, at any month, will be smaller than a risk level α.

5.3 Readiness Evaluation

Many papers have been written during the years on the readiness of operational units. A survey of several approaches to evaluate readiness is given in the paper of Barzily, Marlow, and Zacks (1979). In that paper about 30 references are provided for naval logistics environments. Three approaches are identified: data analysis, theoretical models, and readiness indexes. In the present section we discuss statistical analysis of high dimensional data sets of hierarchically structured binary variables with missing data. This is based on the paper of Zacks, Marlow, and Breir (1985). This analysis was applied to Marine Corps Combat Readiness Evaluation System (MCCRES).

5.3.1 Marine Corps Combat Readiness Evaluation

The Marine Corps Combat Readiness Evaluation System (MCCRES) measures the effectiveness of Marine Corps units, including infantry units (battalions), against well-defined Mission Performance Standards (MPS) in simulated combat environments. The evaluation system for infantry battalions is comprised of 17 MPSs, which are classified into four sections. Section A contains three MPSs, Section B contains seven MPSs, Section C contains four MPSs, and Section D contains three MPSs. In all cases, an MPS consists of several tasks, and each task consists of several requirements. Each requirement is assigned by a team of judges values 1 or 0, according to whether the performance of the requirement is satisfactory or not. An N/A was assigned if a requirement was not applicable (evaluated). A weighted average of all scores in applicable requirements was then calculated, with prescribed weights for each requirement, which were prepared ahead by teams of experts. Thus, each task is assigned these weighted averages, which are called *task-scores*. The MCCRES computes hierarchically a weighted average of all task scores in an MPS to yield an *MPS-score*. All MPS-scores in a section yields a Section-score, and a weighted average of all section scores is a *total-evaluation score* (TES). All scores are based only on applicable evaluations. Accordingly, each evaluation is summarized by 17 vectors of task-scores; 4 vectors of MPS-scores; 1 vector of section-scores, and a TES. Different vectors might have different dimensions, and might include missing components, for the N/As. Task scores might be over-valued or under-valued, due to the missing data (N/As). We show below a method of estimating the basic parameters. The statistical problem is that of analyzing multivariate binary data sets having an extensive amount of incomplete or missing data.

5.3.2   Data Structure

Let t_i denote the number of tasks in the i-th MPS of a section. Let j ($j = 1, \ldots, t_i$) denote the index of the j-th task in the i-th MPS. Let r_{ij} denote the number of requirements in (i, j)-task. Let $k, k = 1, \ldots, r_{ij}$ denote the k-th requirement in task (i, j). Finally, let $n = 1, 2, \ldots, N$, denote the index of the n-th evaluation. We introduce incidental variable $Z_{ijk}^{(n)} = I$ (if requirement i, j, k is applicable in the n-th evaluation), and the random variable $J_{ijk}^{(n)} = I$ (if requirement i, j, k is satisfied in the n-th evaluation). Let $w_{ijk} \geq 0$ be the weight assigned to requirement (i, j, k). $\sum_{k=1}^{r_{ij}} w_{ijk} = 1$, for all (i, j). Moreover, let $w_{ij}^{(n)} = \sum_{k=1}^{r_{ij}} Z_{ijk}^{(n)} w_{ijk}$. Then, the task-score of the n-th evaluation is

$$X_{ij}^{(n)} = \frac{\sum_{k=1}^{r_{ij}} Z_{ijk}^{(n)} J_{ijk}^{(n)} w_{ijk}}{w_{ij}^{(n)}}. \tag{5.22}$$

5.3.3   The Statistical Model

Let $\theta_{ijk}^{(n)}$ denote the probability that requirement (i, j, k) is satisfied in the n-th evaluation. We assume that, for a given (i, j, k), the random variables $\{J_{ijk}^{(n)}, n = 1, \ldots, N\}$ are conditionally independent, given $\{\theta_{ijk}^{(n)}, n = 1, 2, \ldots, N\}$, and that $\{\theta_{ijk}^{(n)}, n = 1, 2, \ldots, N\}$ are independent random variables having some unknown prior distribution, with mean $E\left[\theta_{ijk}^{(n)}\right] = \varphi_{ijk}$. The expected complete task score (ECTS) for a randomly selected evaluation is

$$\tau_{ij} = \sum_{k=1}^{r_{ij}} w_{ijk} \varphi_{ijk}. \tag{5.23}$$

The covariances of $J_{ijk}^{(n)}$ and $J_{ijk'}^{(n)}$ in a randomly chosen evaluation are

$$COV\left(J_{ijk}^{(n)}, J_{ijk'}^{(n)}\right) = I\{k = k'\}\varphi_{ijk}(1 - \varphi_{ijk}) + I\{k \neq k'\}(P_{ij}(k, k') - \varphi_{ijk}\varphi_{ijk'}),$$
$$\tag{5.24}$$

where

$$P_{ij}(k, k') = P\left[J_{ijk}^{(n)} J_{ijk'} = 1\right]. \tag{5.25}$$

The covariance matrix of requirements within the (i, j) task is denoted by (V_{ij}). The variance of a complete task score is

$$\sigma_{ij} = w'_{ij}(V_{ij})w_{ij} \tag{5.26}$$

in which $w'_{ij} = (w_{ij1}, \ldots, w_{ijr_{ij}})$. We are interested also in the correlations between two complete task scores within the i-th MPS. Let $C_{ijj'}(k, l) = COV\left(J_{ijk}^{(n)}, J_{ij'l}^{(n)}\right) = P\left[J_{ijk}^{(n)} J_{ij'l}^{(n)} = 1\right] - \varphi_{ijk}\varphi_{ij'l}$ and $(K_{ijj'})$ a corresponding $r_{ij} x r_{ij'}$ covariance matrix. Then the covariance between two task scores in the same MPS is

$$C_i(j, j') = w'_{ij}\left(K_{ijj'}\right) w_{ij'}. \tag{5.27}$$

Estimators of these covariances, which are based on cases with both requirements available, might yield matrices which are not positive definite, due to missing data. We show in the next section pairwise maximum likelihood estimators, which yield positive definite matrices.

5.3.4 Pairwise Maximum Likelihood Estimators (PMLE)

Consider any two requirements (i, j, k) and (i, j', l) in the same MPS. The likelihood function will be based on the following data for the given two requirements. In order to simplify notation, we will denote the corresponding variables by, $Z_1^{(n)}$, $J_1^{(n)}$, $Z_2^{(n)}$, $J_2^{(n)}$, $n = 1, 2, \ldots, N$. The frequency distribution among the N evaluations of these two variables is shown in Table 5.2.

Table 5.2 Frequency distribution of (Z_1, J_1) and (Z_2, J_2)

Z_1		$Z_2 = 1$		$Z_2 = 0$
1	J_1	$J_2 = 0$	$J_2 = 1$	
	0	f_{00}	f_{01}	$g_{0.}$
	1	f_{10}	f_{11}	$g_{1.}$
0		$g_{.0}$	$g_{.1}$	h

Formally,

$$f_{00} = \sum_{n=1}^{N} Z_1^{(n)} Z_2^{(n)} \left(1 - J_1^{(n)}\right) \left(1 - J_2^{(n)}\right), \tag{5.28}$$

$$f_{10} = \sum_{n=1}^{N} Z_1^{(n)} Z_2^{(n)} J_1^{(n)} \left(1 - J_2^{(n)}\right),$$

$$f_{01} = \sum_{n=1}^{N} Z_1^{(n)} Z_2^{(n)} \left(1 - J_1^{(n)}\right) J_2^{(n)},$$

$$f_{11} = \sum_{n=1}^{N} Z_1^{(n)} Z_2^{(n)} J_1^{(n)} J_2^{(n)},$$

$$g_{0.} = \sum_{n=1}^{N} Z_1^{(n)} \left(1 - Z_2^{(n)}\right) \left(1 - J_1^{(n)}\right),$$

$$g_{1.} = \sum_{n=1}^{N} Z_1^{(n)} \left(1 - Z_2^{(n)}\right) \left(J_1^{(n)}\right),$$

$$g_{.0} = \sum_{n=1}^{N} \left(1 - Z_1^{(n)}\right) Z_2^{(n)} \left(1 - J_2^{(n)}\right),$$

$$g_{.1} = \sum_{n=1}^{N} \left(1 - Z_1^{(n)}\right) Z_2^{(n)} J_2^{(n)},$$

$$h = \sum_{n=1}^{N} \left(1 - Z_1^{(n)}\right) \left(1 - Z_2^{(n)}\right).$$

Let $M = f_{00} + f_{01} + f_{10} + f_{11}$. Furthermore, $N = M + g_{0.} + g_{1.} + g_{.0} + g_{.1} + h$. The g frequencies are for partially observed data, and h is the number of evaluations in which both requirements are unavailable. The cell probabilities are

$$\alpha = P\left[J_1^{(n)} = 0, J_2^{(n)} = 0\right], \tag{5.29}$$

$$\beta = P\left[J_1^{(n)} = 0, J_2^{(n)} = 1\right],$$

$$\gamma = P\left[J_1^{(n)} = 1, J_2^{(n)} = 0\right].$$

The kernel of the log-likelihood is

$$l(\alpha, \beta, \gamma) = f_{00} \log(\alpha) + f_{01} \log(\beta) + f_{10} \log(\gamma) + f_{11} \log(1 - \alpha - \beta - \gamma)$$
$$+ g_{0.} \log(\alpha + \beta) + g_{1.} \log(1 - \alpha - \beta) + g_{.0} \log(\alpha + \gamma)$$
$$+ g_{.1} \log(1 - \alpha - \gamma). \tag{5.30}$$

Let $\hat{\theta} = (\alpha, \beta, \gamma)$. The point $\hat{\theta}$ in the simplex $\Theta = \{0 \leq \alpha, \beta, \gamma \leq 1 : \alpha + \beta + \gamma = 1\}$, which is maximizing (5.30), is called pair-wise MLE (PMLE). We apply the E-M algorithm of Dempster, Laird, and Rubin (1977) for obtaining the PMLEs iteratively. The E-M algorithm consists of iterative cycles of two stages. In the first stage of a cycle (called the M-stage) one computes the estimates

$$\hat{\alpha}^{(p)} = \frac{f_{00}^{(p)}}{N}, \quad p = 0, 1, 2, \ldots \quad \hat{\beta}^{(p)} = \frac{f_{01}^{(p)}}{N}, \quad \hat{\gamma}^{(p)} = \frac{f_{10}^{(p)}}{N} \tag{5.31}$$

starting with $f_{ij}^{(0)} = \frac{f_{ij} + 0.5}{M + 2}$.

In the second stage of a cycle (called the E-stage) one computes, the conditional expectations of the frequencies, namely

$$f_{00}^{(p+1)} = f_{00} + \frac{g_{0.}\hat{\alpha}^{(p)}}{\hat{\alpha}^{(p)} + \hat{\beta}^{(p)}} + \frac{g_{.0}\hat{\alpha}^{(p)}}{\hat{\alpha}^{(p)} + \hat{\gamma}^{(p)}} + h\hat{\alpha}^{(p)}, \tag{5.32}$$

$$f_{01}^{(p+1)} = f_{01} + \frac{g_{0.}\hat{\beta}^{(p)}}{\hat{\alpha}^{(p)} + \hat{\beta}^{(p)}} + \frac{g_{.1}\hat{\beta}^{(p)}}{1 - \hat{\alpha}^{(p)} - \hat{\gamma}^{(p)}} + h\hat{\beta}^{(p)},$$

$$f_{10}^{(p+1)} = f_{10} + \frac{g_{1.}\hat{\gamma}^{(p)}}{1 - \hat{\alpha}^{(p)} - \hat{\beta}^{(p)}) + g_{.0}\hat{\gamma}^{(p)}}(\hat{\alpha}^{(p)} + \hat{\gamma}^{(p)}) + h\hat{\gamma}^{(p)}.$$

Return to stage 1 of the $(p + 1)$th cycle.

In Tables 5.3 and 5.4 we present an example of such estimation.

We see that in this example, convergence to the solution (fixed point) was in about five iterations.

The asymptotic (large N) variance–covariance matrix of the PMLE is the inverse of the Fisher Information matrix (FIM). Let $M_{1.} = g_{0.} + g_{1.}$, and $M_{.1} = g_{.0} + g_{.1}$. The partial derivatives of the log-likelihood (5.30) are called the score-functions. These are

Table 5.3 Frequency distribution

		$Z_2 = 1$		
Z_1	J_1	$J_2 = 0$	$J_2 = 1$	$Z_2 = 0$
1	0	20	10	5
	1	5	15	5
0		5	5	5

Table 5.4 Iterative PMLE

p	$\alpha^{(p)}$	$\beta^{(p)}$	$\gamma^{(p)}$
0	0.3942	0.2019	0.1058
1	0.3896	0.1963	0.1053
2	0.3894	0.1947	0.1048
3	0.3896	0.1940	0.1045
4	0.3897	0.1940	0.1045
5	0.3897	0.1940	0.1045

$$
\frac{\partial}{\partial \alpha} l(\alpha, \beta, \gamma) = \frac{f_{00}}{\alpha} - \frac{f_{11}}{1 - \alpha\beta\gamma}
$$

$$
+ \frac{g_{0.}}{(\alpha\beta)(1 - \alpha\beta)} - \frac{M_{1.}}{1 - \alpha\beta} + \frac{g_{\cdot 0}}{(\alpha + \gamma)(1 - \alpha - \gamma)} - \frac{M_{.1}}{1 - \alpha\gamma},
$$

$$
\frac{\partial}{\partial \beta} l(\alpha, \beta, \gamma) = \frac{f_{01}}{\beta} - \frac{f_{11}}{1 - \alpha\beta\gamma} + \frac{g_{0.}}{(\alpha + \beta)(1 - \alpha - \beta)} - \frac{M_{1.}}{1 - \alpha - \beta},
$$

$$
\frac{\partial}{\partial \gamma} l(\alpha, \beta, \gamma) = \frac{f_{10}}{\gamma} - \frac{f_{11}}{1 - \alpha - \beta - \gamma} + \frac{g_{.0}}{(\alpha + \gamma)(1 - \alpha - \gamma)} - \frac{M_{.1}}{1 - \alpha\gamma}.
$$

$$(5.33)$$

The FIM is the variance–covariance matrix of the gradient vector of the log-likelihood, i.e.,

$$
I(\theta) = M \begin{pmatrix} A\left(\frac{1-\beta-\gamma}{\alpha}\right) + b + C & A + B & A + C \\ A + B & A\left(\frac{1-\alpha-\gamma}{\beta}\right) + B & A \\ A + C & A & A\left(\frac{1-\alpha-\beta}{\gamma}\right) + C, \end{pmatrix} \quad (5.34)
$$

where

$$
A = \frac{1}{1 - \alpha - \beta - \gamma},
$$

$$
B = \frac{M_{1.}}{M(\alpha + \beta)(1 - \alpha - \beta)} \quad (5.35)
$$

$$
C = \frac{M_{.1}}{M(\alpha + \gamma)(1 - \alpha - \gamma)}.
$$

The FIM and its inverse can be easily computed.

5.4 Empirical Bayes Applied to MCCRES Data

The present section is based on the paper of Breir, Zacks, and Marlow (1986). Reduction of dimensionality in the present study was by classifying all 907 requirements to 6 categories, which are:

1. Advanced preparation and education.
2. Combat service support and logistics.
3. Mission planning and preparation.
4. Command control and task organization.
5. Execution.
6. Information and Communication.

The data set on which the study consists of $N = 98$ evaluation. The objective is to develop efficient estimators of the vectors of $K = 6$ probabilities of satisfying the requirements in the categories. We develop empirical Bayes estimators (EBE) as well as Stein-type estimators (STE) of these vectors, and compare their efficiencies. See the article of James and Stein (1960).

In typical evaluation only about 50% of the requirements were tested. The number of tested requirements in the i-th evaluation and the j-th category is denoted by M_{ij}. Let X_{ij} denote the number of successfully requirement among the tested ones. The data are based on the statistics $\{(X_{ij}, M_{ij}), i = 1, \ldots, N; j = 1, \ldots, K\}$. The basic assumption of the statistical model is that the binary (0,1) responses J_{ij} on those requirements in the same category (1 for satisfied and 0 for not satisfied) are independent random variables, having the same probability of 1, which is $\theta_{ij} = P\{J_{ij} = 1\}$. This implies that $X_{ij}|\theta_{ij} \sim Binom(M_{ij}, \theta_{ij})$. We apply the variance stabilizing transformation

$$Y_{ij} = 2 \sin^{-1} \left(\frac{X_{ij+3/8}}{M_{ij+3/4}} \right)^{1/2} \quad i = 1, \ldots, N; j = 1, \ldots, k. \quad (5.36)$$

It is well known Johnson and Kotz (1969) that for large values of M_{ij} the distribution of Y_{ij} is approximately normal , $N(\eta_{ij}, 1/M_{ij})$, where $\eta_{ij} = 2 \sin^{-1}(\theta_{ij})^{1/2}$.

5.4.1 Stein-Type Estimators

James and Stein (1960) proved that if $\mathbf{Y} \sim \mathbf{N_p}(\boldsymbol{\eta}, \mathbf{V})$ is a p-variate multi-normal random vector, and $p \geq 3$, then the estimator of η

$$\hat{\boldsymbol{\eta}} = \left(1 - \frac{\mathbf{p} - 2}{\mathbf{Y'V^{-1}Y}} \right) \mathbf{Y}, \quad (5.37)$$

with the risk function $E\left[\|\hat{\eta} - \eta\|^2\right]$, has uniformly smaller risk than the MLE. We consider modified estimators,

$$hat\eta_{\mathbf{i}} = \bar{Y}_i 1_{\mathbf{k}} + \left(1 - \frac{k - 2}{\sum_{j=1}^k M_{ij}(Y_{ij} - \bar{Y}_i)^2}\right)^+ (\mathbf{Y}_i - \bar{Y}_i 1_k), \qquad (5.38)$$

where $\bar{Y}_i = \sum_{j=1}^k M_{ij} Y_{ij} / \sum_{i=1}^k M_{ij}$, and 1_k is a k-dimensional vector of 1's. A similar estimator was considered by Morris (1983). Finally, every component of $\hat{\boldsymbol{\eta}}_{\mathbf{i}}$ is transformed to yield

$$\hat{\theta}_{ij}^{(ST)} = \sin^2(\hat{\eta}_{ij}/2), i = 1, \ldots N; j = 1, \ldots, k. \qquad (5.39)$$

5.4.2 Empirical Bayes Estimators

The Bayesian model is represented by the following assumptions:

1. The conditional distribution of $\mathbf{Y}_i$, given $\boldsymbol{\eta}_i$, $(i = 1, 2, \ldots, N)$, is normal with mean η and a known covariance matrix $D_i = \text{diag}\,[1/M_{i1}, \ldots, 1/M_{ik}]$, i.e., $\mathbf{Y}_i | \boldsymbol{\eta}_i \sim N(\boldsymbol{\eta}_i, D_i)$.
2. $\mathbf{Y}_1, \ldots, \mathbf{Y}_N$ are independent.
3. $\boldsymbol{\eta}_1, \ldots, \boldsymbol{\eta}_N$ are independent identically distributed vectors, having a prior normal distribution $N(\mu, \varXi)$.

The empirical Bayes approach is to estimate the prior parameters $(\mu, \varXi)$ consistently from the data. See also the paper of Zacks, Marlow, and Breir (1985).

The posterior distribution of $\boldsymbol{\eta}_i$, given $\{\mathbf{Y}_i, i = 1, \ldots, N\}$, is normal with posterior mean

$$E\left[\boldsymbol{\eta}_i | \mathbf{Y}_i, \mu, \varXi\right] = \mathbf{Y}_i - B_i(\mathbf{Y}_i - \mu), \qquad (5.40)$$

and posterior covariance matrix $(I - B_i)D_i$, where

$$B_i = D_i(D_i + \varXi)^{-1}. \qquad (5.41)$$

Thus, for specified $(\mu, \varXi)$, the Bayesian estimator of $\boldsymbol{\eta}_i$, for the squared error loss, is $\hat{\boldsymbol{\eta}}_i = \mathbf{Y}_i - B_i(\mathbf{Y}_i - \mu)$.

The log-likelihood function of $(\mu, \varXi)$, given the sample data, is

$$l(\mu, \varXi) = -\frac{1}{2}\sum_{i=1}^N \log |\varXi + D_i| - \frac{1}{2}\sum_{i=1}^N (\mathbf{Y}_i - \mu)'(\varXi + D_i)^{-1}(\mathbf{Y}_i - \mu). \qquad (5.42)$$

The vector $\hat{\boldsymbol{\mu}}(\boldsymbol{\varXi})$ which maximizes (5.42), when $\boldsymbol{\varXi}$ is known, is

$$\hat{\boldsymbol{\mu}}(\boldsymbol{\varXi}) = (\sum_{i=1}^{N}(\boldsymbol{\varXi} + D_i)^{-1})^{-1} \sum_{i=1}^{N}(\boldsymbol{\varXi} + D_i)^{-1}\mathbf{Y}_i. \tag{5.43}$$

The matrix $\hat{\boldsymbol{\varXi}}$ is an MLE matrix, maximizing (5.42). Accordingly, $\hat{\boldsymbol{\mu}}(\hat{\boldsymbol{\varXi}})$ is an MLE of $\boldsymbol{\mu}$. We apply the E-M algorithm to converge to $\hat{\boldsymbol{\mu}}(\hat{\boldsymbol{\varXi}})$. The algorithm proceeds in two phases: from phase E to phase M, and back to phase E, until convergence is attained. Notice first that if $\boldsymbol{\eta}_i i = 1, \ldots, N$ were known, the kernel of the log-likelihood of $(\boldsymbol{\mu}, \boldsymbol{\varXi})$ would have been

$$l^*(\boldsymbol{\mu}, \boldsymbol{\varXi}) = -\frac{N}{2} \log |\boldsymbol{\varXi}| - \frac{1}{2} \sum_{i=1}^{N}(\boldsymbol{\eta}_i - \boldsymbol{\mu})'\boldsymbol{\varXi}^{-1}(\boldsymbol{\eta}_i - \boldsymbol{\mu}). \tag{5.44}$$

The E-M algorithm considers the $\boldsymbol{\eta}_i$ vectors as missing data.

The two phases of the algorithm can be specified, after the p-th iteration, as follows:

(I) Phase E:
   Determine

$$Q^{(p)}(\boldsymbol{\mu}, \boldsymbol{\varXi}) = -\frac{1}{2}E\left[\sum_{i=1}^{N}(\boldsymbol{\eta}_i - \boldsymbol{\mu})'\boldsymbol{\varXi}(\boldsymbol{\eta}_i - \boldsymbol{\mu})|(\mathbf{Y}), \boldsymbol{\mu}^{(p)}, \boldsymbol{\varXi}^{(p)}\right] - \frac{N}{2}\log|\boldsymbol{\varXi}|,$$
$$\tag{5.45}$$

where $(\mathbf{Y}) = (\mathbf{Y}_1, \mathbf{Y}_2, \ldots, \mathbf{Y}_N)$, and $\boldsymbol{\mu}^{(p)}, \boldsymbol{\varXi}^{(p)}$ are values of $\boldsymbol{\mu}, \boldsymbol{\varXi}$ determined in the M Phase of the p-th iteration.

(II) Phase M:
   Determine $\boldsymbol{\mu}^{(p+1)}$ and $\boldsymbol{\varXi}^{(p+1)}$ which maximize $Q^{(p+1)}(\boldsymbol{\mu}, \boldsymbol{\varXi})$. Notice that the conditional expectation in (5.45) is that of the conditional normal distribution of $\boldsymbol{\eta}_i$, given $\mathbf{Y}_i$, $\boldsymbol{\mu} = \boldsymbol{\mu}^{(p)}$, $\boldsymbol{\varXi} = \boldsymbol{\varXi}^{(p)}$. This posterior distribution is denoted by $N(\mathbf{w}_i^{(p)}, V_i^{(p)})$, where

$$\mathbf{w}_i^{(p)} = (I - B_i^{(p)})\mathbf{Y_i} + B_i^{(p)}\boldsymbol{\mu}^{(p)} \tag{5.46}$$

and

$$V_i^{(p)} = B_i^{(p)}\boldsymbol{\varXi}^{(p)}. \tag{5.47}$$

Let $\bar{V}^{(p)} = \frac{1}{N}\sum_{i=1}^{N}V_i^{(p)}$, then we obtain that

$$Q^{(p)}(\boldsymbol{\mu}, \boldsymbol{\varXi}) = -\frac{N}{2}\log|\boldsymbol{\varXi}| - \frac{N}{2}tr\left[\boldsymbol{\varXi}^{-1}\bar{V}^{(p)}\right]$$

$$-\frac{1}{2}\sum_{i=1}^{N}(\mathbf{w}_i^{(p)} - \boldsymbol{\mu})'\boldsymbol{\varXi}^{-1}\left(\mathbf{w}_i^{(p)} - \boldsymbol{\mu}\right). \tag{5.48}$$

Thus, we have to determine $\mu^{(p+1)}$ and $\mathcal{Z}^{(p+1)}$ to maximize (5.48). We immediately obtain that

$$\mu^{(p+1)} = \frac{1}{N} \sum_{i=1}^{N} \mathbf{w}_i^{(p)}. \tag{5.49}$$

Substituting (5.49) in (5.48) we get

$$Q^{(p)}(\mu^{(p+1)}, \mathcal{Z}) = -\frac{N}{2} \log |\mathcal{Z}| - \frac{N}{2} tr \left[\mathcal{Z}^{-1} \left(C^{(p)} + \bar{V}^{(p)} \right) \right], \tag{5.50}$$

where $C^{(p)}$ is a $k x k$ covariance matrix of the rows of the matrix $W^{(p)} = (\mathbf{w}_1, \mathbf{w}_2, \ldots, \mathbf{w}_N)$, i.e.,

$$C^{(p)} = W^{(p)} \left(I_N - \frac{1}{N} 1_N 1_N' \right) \left(W^{(p)} \right)'. \tag{5.51}$$

Finally, the matrix $\mathcal{Z}^{(p+1)}$ which maximizes (5.50) is

$$Xi^{(p+1)} = C^{(p)} + \bar{V}^{(p)}. \tag{5.52}$$

In Table 5.5, we present the results of $\mu^{(p)}$ and the correlation matrix $R^{(p)}$ corresponding to $\mathcal{Z}^{(p)}$. We started with $\mu^{(0)} = \bar{Y}_N$, and $\mathcal{Z}^{(0)} =$

Table 5.5 $\mu^{(p)}$ and $R^{(p)}$ in the E-M algorithm, based on 98 evaluations

p	$\mu^{(p)}$	1	2	3	4	5	6
1	2.4655	1.0000	0.8598	0.8079	0.7561	0.7458	0.7790
	2.5330		1.0000	0.8147	0.7859	0.7759	0.8366
	2.3935			1.0000	0.8567	0.9196	0.8530
	2.5268				1.0000	0.7697	0.8684
	2.3445					1.0000	0.7777
	2.5031						1.0000
9	2.4662	1.0000	0.8775	0.8171	0.7617	0.7548	0.7806
	2.5341		1.0000	0.8395	0.8008	0.7990	0.8371
	2.3937			1.0000	0.8698	0.9228	0.8570
	2.5262				1.0000	0.7733	0.8740
	2.3445					1.0000	0.7761
	2.5036						1.0000
10	2.4662	1.0000	0.8792	0.8175	0.7618	0.7550	0.7808
	2.5341		1.0000	0.8414	0.8019	0.8005	0.8373
	2.3937			1.0000	0.8705	0.9231	0.8572
	2.5262				1.0000	0.7732	0.8741
	2.3445					1.0000	0.7760
	2.5036						1.0000

$S - \text{diag}\left[1/\bar{M}_1, \ldots, 1/\bar{M}_k\right]$, where $S = \frac{1}{N}\sum_{i=1}^{N}\left(\mathbf{Y}_i\mathbf{Y}_i' - \bar{\mathbf{Y}}_N\bar{\mathbf{Y}}_N'\right)$, and $\bar{\mathbf{Y}}_N = \frac{1}{N}\sum_{i=1}^{N}\mathbf{Y}_i$.

We see in this numerical example that $\boldsymbol{\mu}^{(p)}$ converges to the MLE quite rapidly. The convergence of $R^{(p)}$ is somewhat slower. Ten iterations provide stable estimates of $R^{(p)}$ and therefore of $\Xi^{(p)}$. The empirical Bayes estimators of the success probabilities θ_{ij} are obtained from the corresponding EBE of η_{ij} according to the formula

$$\theta_{ij}^{(EB)} = \sin^2\left(\hat{\eta}_{ij}(\hat{\boldsymbol{\mu}}, \hat{\Xi})/2\right). \tag{5.53}$$

5.4.3　Numerical Comparison of the Estimators

We compare first the STE and the EBE estimates of the success probabilities to those of the MLE estimators, which are $\theta_{ij}^{(ML)} = X_{ij}/M_{ij}$. All estimates are based on the MACCRES data. In Table 5.6 we present the estimates for two evaluations.

We compare the efficiency of the MLE and SET, relative to the EBE, by cross-validation, which has been used to test the validity of a regression model on new data. Accordingly, we partition the 907 requirements, in all the 98 evaluations, into two parts of almost equal size. We label the parts P^A and P^B. The first part P^A is then used to determine $\theta_{ij.A}^{(ML)}$, $\theta_{ij.A}^{(ST)}$, and $\theta_{ij.A}^{EB)}$. In order to improve somewhat the performance of the MLE, we modify it to be $\theta_{ij.A}^{(ML)} = (X_{ij}^{(A)} + 0.5)/(M_{ij}^{(A)} + 1)$. The above three estimators from part A were compared to $P_{ij} = \theta_{ij.B}^{(ML)}$ which are the modified proportions of success. Two discrepancy indexes were used. The first one is

$$MD^{(.)} = \sum_{i=1}^{N}\sum_{j=1}^{k}\left[\frac{M_{ij}^{(B)}\left(\theta_{ij.A}^{(.)} - P_{ij}\right)^2}{P_{ij}\left(1 - P_{ij}\right)}\right]. \tag{5.54}$$

Table 5.6 Estimates of success probabilities in the six categories

Eval	Est.	Cat. 1	Cat. 2	Cat. 3	Cat. 4	Cat. 5	Cat. 6
5	MLE	1.000	1.000	1.000	0.915	0.953	0.859
	STE	0.981	0.973	0.980	0.918	0.945	0.875
	EBE	0.973	0.957	0.958	0.925	0.950	0.904
12	MLE	0.911	0.957	0.915	0.945	0.741	0.918
	STE	0.901	0.929	0.904	0.933	0.775	0.909
	EBE	0.899	0.916	0.845	0.934	0.786	0.914

Table 5.7 Comparison of
EBE, STE, and MLE by
cross-validation

Partition	Estimator	$MD^{(.)}$	$L^{(.)}$
1	EBE	1072.62	369.01
	STE	1263.64	427.31
	MLE	1596.94	522.87
2	EBE	999.02	332.91
	STE	1163.12	392.19
	MLE	1452.34	475.27
3	EBE	1028.61	351.46
	STE	1183.41	410.51
	MLE	1600.51	517.38

This is the Mahalanobis squared distance between P^A and P^B. Formula (5.54) has also the form of Pearson chi-squared statistic. The second measure of discrepancy is

$$
L^{(.)} = \sum_{i=1}^{N} \sum_{j=1}^{k} \left[M_{ij}^{(B)} P_{ij} \log \left(\frac{P_{ij}}{\theta_{ij.A}^{(.)}} \right) + M_{ij}^{(B)} \left(1 - P_{ij} \right) \log \left(\frac{1 - P_{ij}}{1 - \theta_{ij.A}^{(.)}} \right) \right].
$$
(5.55)

This statistic is similar to the Kullback–Leibler information (see Zacks 2014, p. 206). In Table 5.7 we present these measures of discrimination for three evaluations.

As seen in Table 5.7, the minimal discrimination statistic is always that of the EBE. We can thus measure the relative-efficiency of the STE and MLE compared to the EBE. For example, for Partition 1 and $MD^{(.)}$, the relative efficiency of STE is $1072.62/1263.64 = 0.849$; and with respect to $L^{(.)}$ it is $369.01/427.31 = 0.864$. Similar results we get also in other partitions. The MLE is even less efficient than the STE compared to EBE. Its relative efficiency for partition 1 and $MD^{(.)}$ is $1072.62/1596.94 = 0.672$. See also the paper of Zacks (1988), on simulation study using empirical Bayes techniques.

5.5   Time of Entrance to the Wear-Out Phase

The problem discussed in the present section is that of detecting the time a system enters wear-out phase. In reliability theory, the haza wear-out phase.ard (failure rate) function of systems is described as a U-shaped one. At the very early life of a system, it often has a decreasing failure rate function. This period in the life of a system is called the "infant mortality" phase. During this phase systems are debugged for all possible causes for failures, or parts are replaced. Systems which overcome this first phase usually enter a second phase of constant, low, hazard. This is a phase in which the inter-failure times are exponentially distributed. Electronic systems often remain in this stable phase until they completely fail. Mechanical systems often enter after a while a third phase in which the failure rate starts to increase monotonically. This phase is called a wear-out phase. The problem in maintenance is to determine whether a system is already in this third phase. In the early 1980s, when I was consulting for the Program in Logistics, we were approached by officers from Naval Aviation with a problem of this nature. They were training new pilots on old Phantom jets, and they realized that the times between failures of engines became shorter than in the early age of these airplanes. The question was whether their engines are already in the wear-out phase. I published a paper on this subject, which will be discussed below (see Zacks 1984).

5.5.1 The Exponential-Weibull Wear-Out Life Distribution

The family of distribution considered here is based on the following nondecreasing failure rate function

$$h\left(t; \lambda, \alpha, \tau\right) = I\left[0 < t \leq \tau\right]\lambda + I\left[\tau < t\right]\left(\lambda + \lambda^{\alpha}\alpha\left(t - \tau\right)^{\alpha-1}\right). \tag{5.56}$$

The parameter τ is the time of change from exponential distribution to a Weibull distribution. The problem is to estimate the change-point τ on the basis of data up to time t, and to decide whether $[t \leq \tau]$ or $[t > \tau]$. From the failure rate function (5.56) we derive the Exponential-Weibull distribution, which is

$$f(x, \lambda, \alpha, \tau) = \lambda e^{-\lambda x}(-(\lambda(x-t)^{+})^{\alpha}. \tag{5.57}$$

Obviously, $F(x; \lambda, \alpha, \tau) = 0$ for all $x \leq 0$. The corresponding density function is

$$f(x; \lambda, \alpha, \tau) = \lambda e^{-\lambda x}\left(1 + \alpha\left(\lambda\left(x-\tau\right)^{+}\right)^{\alpha-1}\right)\exp\left[-\lambda^{\alpha}\left(x-\tau\right)^{+\alpha}\right], x \geq 0. \tag{5.58}$$

It follows that the residual life distribution, in the wear-out phase, is

$$g\left(x-\tau; \lambda, \alpha\right) = \lambda e^{-\lambda(x-\tau)}\left(1 + \alpha\left(\lambda\left(x-\tau\right)\right)^{\alpha-1}\right)\exp\left[-\lambda^{\alpha}\left(x-\tau\right)^{\alpha}\right], x \geq \tau. \tag{5.59}$$

We present now the moments of the E-W distribution, with $\lambda = 1$, namely $\mu_{r}(\alpha, \tau)$. If $\lambda \neq 1$, then

$$E\left[X^{r}; \lambda, \alpha, \tau\right] = \left(\frac{1}{\lambda^{r}}\right)\mu_{r}(\alpha, \lambda\tau).$$

Table 5.8 First four
moments of the E-W
distribution, $\lambda = 1, \alpha = 2$

r	$\tau = 0$	$\tau = 0.5$	$\tau = 1$	$\tau = 1.5$
1	0.5456	0.7244	0.8328	0.8986
2	0.4544	0.7869	1.0971	1.3510
3	0.4777	1.0376	1.7612	2.5315
4	0.5903	1.5584	3.1816	5.3489

For $\tau > 0$,

$$\mu_r(\alpha, \tau) = \int_0^\tau x^r e^{-x} dx + \int_\tau^\infty x^r e^{-x-(x-\tau)^\alpha} (1 + \alpha(x - \tau)^{\alpha-1}) dx$$

$$= r!(1 - Pos(r|\tau)) + e^{-r} \sum_{j=0}^r \binom{r}{j} \tau^{r-j} M_j(\alpha), \tag{5.60}$$

where

$$M_j(\alpha) = j \int_0^\infty y^{j-1} e^{-(y+y^\alpha)} dy, \tag{5.61}$$

and $Pos(r|\tau)$ is the c.d.f. of the Poisson distribution at r with mean τ.

In Table 5.8 we present some values of these moments.

5.5.2 Bayesian Estimates of τ

In this section we develop a Bayesian framework for estimating the change point τ. Given a sequence of replacement times, $0 < t_1 < t_2, \ldots$ we consider the inter-replacement random variables $X_i = t_i - t_{i-1}$, where $t_0 = 0$. The preventive maintenance replaces parts, even before failures, if the inter-replacement time is greater than Δ. Thus, $X_i = \min[\Delta, Y_i]$, where $Y_1, Y_2, \ldots$ are independent random variables, having an E-W distribution, with parameters $\lambda, \alpha, \tau_i = (\tau - t_i)^+$. The data is right-censored. Let $J_i = I[X_i < \Delta]$ and let $\mathbf{X}^{(n)} = (X_1, \ldots, X_n)$. The likelihood of τ is

$$L\left(\tau; \mathbf{X}^{(n)}\right) = I\{\tau = 0\} \exp\left[-\sum_{i=1}^n X_i^\alpha\right] \prod_{i=1}^n \left(1 + \alpha J_i X_i^{\alpha-1}\right)$$

$$+ \sum_{j=1}^{n-1} I\left[t_{j-1} < \tau \le t_j\right] \exp\left[-\sum_{k=j+1}^n X_k^\alpha\right] \prod_{k=j+a}^n \left(1 + \alpha J_k X_k^{\alpha-1}\right)$$

$$\times \left(1 + \alpha J_j (X_j + t_{j-1} - \tau)^{\alpha-1}\right) \exp\left[-(X_j + t_{j-1} - \tau)^{\alpha-1}\right]$$

$$\times I\left[t_{n-1} < \tau \le t_n\right] \exp\left[-(X_n + t_{n-1} - \tau)^\alpha\right]$$

$$\times \left(1 + \alpha J_n (X_n + t_{n-1} - \tau)^{\alpha-1}\right) + I\left[\tau > t_n\right]. \tag{5.62}$$

We adopt the following prior distribution (c.d.f.) for τ

$$\Psi(\tau) = I\,[\tau \geq \tau_0]\,\big[p + (1-p)(1 - \exp(-\psi(\tau - \tau_0)))\big]. \tag{5.63}$$

p is the prior probability that the change-point τ occurred before τ_0. According to the E-W distribution, if a change-point τ has not occurred, the next time for part replacement would have the distribution of the minimum of X and τ_0. The prior distribution (5.63) assumes that the distribution after τ_0 is exponential with mean $\frac{1}{\psi}$.

The posterior c.d.f. of τ,, given $\mathbf{X}^{(n)}$, is for given α and $\lambda = 1$,

$$\Psi(\tau|\mathbf{X}^{(n)}) = I\,[\tau \geq \tau_0]\,\frac{\int_0^\tau L\left(t; X^{(n)}\right) d\Psi(t)}{\int_0^\infty L\left(t; X^{(n)}\right) d\Psi(t)}. \tag{5.64}$$

The posterior probability, given $\mathbf{X}^{(n)}$, that τ has already occurred, is

$$p_n = 1 - \frac{(1-p)e^{-\psi t_n}}{D(p, \psi; \mathbf{X}^{(n)})}, \tag{5.65}$$

where

$$D(p, \psi; \mathbf{X}^{(n)}) = \int_0^\infty L(t; X^{(n)})d\Psi(t) = p \exp\left[-\sum_{i=1}^n X_i^\alpha\right]\prod_{i=1}^n \left(1 + \alpha J_i X_i^{\alpha-1}\right)$$

$$+ (1-p)\left[\sum_{j=1}^n \exp\left[-\sum_{k=j+1}^n X_k^\alpha\right]\prod_{k=j+1}^n \left(1 + \alpha J_k X_k^{\alpha-1}\right)e^{-\psi t_{j-1}}\psi\right.$$

$$\times \int_0^{X_j} e^{-\psi u}\left(1 + \alpha J_j\left(X_j - u\right)^{\alpha-1}\right)e^{-(X_j-u)^\alpha} du + \psi e^{-\psi t_{n-1}}$$

$$\times \int_0^{X_n} e^{-\psi u}\left(1 + \alpha J_n\left(X_n - u\right)^{\alpha-1}\right)$$

$$\left.\times\, e^{-(X_n-u)^\alpha} du + e^{-\psi t_n}\right]. \tag{5.66}$$

If p_n is large we have high evidence that the change τ has already occurred. The Bayesian estimator of τ is

$$\hat{\tau}\left(p, \psi, \mathbf{X}^{(n)}\right) = (1-p)\left[\psi\sum_{j=1}^n \exp\left[\sum_{k=j+1}^n X_k^\alpha\right]\prod_{k=j+1}^n \left(1 + \alpha J_k X_k^{\alpha-1}\right)\right.$$

$$\times\, e^{-\psi t_j}\psi\left[\int_0^{X_j}\left(t_{j-1} + u\right)e^{-\psi u}\left(1 + \alpha J_j\left(X_j - u\right)^{\alpha-1}\right)\right.$$

$$\times \, e^{-(X_j-u)^\alpha} du + \psi e^{-\psi t_{n-1}} \int_0^{X_n} (t_{n-1}+u)\, e^{-\psi u}$$

$$\times \left(1 + \alpha J_n \, (X_n - u)^{\alpha-1}\right) e^{-(X_n-u)^\alpha} du$$

$$+ \, \psi e^{-\psi t_n} \int_0^\infty (t_n + u)\, e^{-\psi u} du \bigg] / D\left(p, \psi; \mathbf{X}^{(n)}\right) \Bigg].$$

$$(5.67)$$

We next present some numerical simulations, which illustrate the theory. We consider a system with $\lambda^{-1} = 200[hr]$, $\Delta = 225[hr]$, $\tau = 750[hr]$, $\alpha = 2$, and $p = 0.2$. The values of p_n and $\hat{\tau}_n$ are computed adaptively after each stage. Table 5.9 presents such simulation run.

We see in Table 5.9 that if our stopping rule is $\min\{n : t_n \geq \hat{\tau}_n\}$, then the estimation error is 116.27. On the other hand, if our stopping rule is $\min\{n : p_n \geq 0.8\}$, then the estimation error is 8.42. These results are based only on one simulation run. To obtain more reliable information, we repeated these simulation runs 100 times, with three values of ψ, and the stopping rule $\min\{n : p_n \geq 0.9\}$. These results are given in Table 5.10.

The values in Table 5.10 indicate that larger values of ψ yield better results.

Table 5.9 Simulation run with parameters $\lambda = 1/200$, $\psi = 1/1000$, $p = 0.2$, $\tau = 750$

n	X_n	J_n	t_n	p_n	$\hat{\tau}_n$
1	41.49	1	41	0.288	742.71
2	156.56	1	198	0.473	646.85
3	108.39	1	306	0.631	514.63
4	225.00	0	531	0.437	921.39
5	225.00	0	756	0.309	1317.66
6	31.19	1	788	0.385	1237.91
7	23.03	1	811	0.446	1170.46
8	115.07	1	926	0.610	1011.46
9	114.04	1	1040	0.742	866.27
10	20.53	1	1060	0.779	815.35
11	27.09	1	1087	0.820	758.42

Table 5.10 Means and standard-errors of $[\hat{\tau}_n - \tau]$ in 100 simulation runs

ψ	1/500	1/750	1/1000
Mean	65.66	152.85	202.13
Std.-Error	311.78	325.40	322.17

Chapter 6
Foundation of Sampling Surveys

There are two basic approaches to sampling from finite populations: The Design Approach and the Modeling Approach. In the Design Approach randomness is introduced into the schemes of choosing the elements (units) of the populations to the samples. In the Modeling Approach the values observed of the population units are considered random variables, having a prescribed joint distributions. This was also called in the literature the super-population approach.

There are several different problems connected with designs of sampling from finite populations. One problem is whether the units should be chosen at random, or by a non-randomized strategy. If by a non-randomized strategy the question is, which units should be selected to the sample? A third problem is, how many units should be sampled? In a Bayesian prediction approach, certain prior assumptions entail that, given the data, the posterior distribution is independent of the sampling rule, and therefore units should not be chosen at random. However, in order to minimize the prediction risk, the optimal sample depends on the available information. For example, if covariates are available for all units of the population, it might be optimal to choose units having the largest covariate(s). How large the sample should be depends on the precision criterion. The larger the sample is, the prediction risk would generally be smaller. Many papers were written on these questions. We mention some general results, which were proved rigorously in the paper of Zacks (1969a). This paper treats the problem of selecting a sample of fixed size n from a finite population, according to the principle of minimizing the expected Bayes risk. It is shown that generally the optimal choice of units is sequential one and non-randomized. This means that after each observation one should evaluate the results and decide which unit to select next. Furthermore, optimal Bayes sampling selection should be without replacements. Considering the results of this paper, the question arises whether there is any role for randomization? A Bayesian may sometimes face very complicated situations, in which the prior information is

© Springer Nature Switzerland AG 2020

S. Zacks, *The Career of a Research Statistician*, Statistics for Industry, Technology, and Engineering, https://doi.org/10.1007/978-3-030-39434-9_6

seriously incomplete. In such cases it might be reasonable to start with a randomized choice of a sample, and then to proceed in a Bayesian fashion. Minimax designs, in cases of incomplete prior information, may also turn to be randomized.

In the present chapter we start with topics related to the Design Approach, and afterwards we discuss the elements of the Modeling Approach. We discuss also Bayesian optimal allocation in stratified random sampling. The results of optimal Bayesian sampling might be good for one choice of prior distribution, but not optimal if the choice of prior is erroneous. This is the problem of robustness of the Bayes procedure against such wrong assumptions. The reader is referred to the paper of Zacks and Solomon (1981) to read more about these problems. The presentation of the next two sections follows the one in the book of Bolfarine and Zacks (1992).

6.1   Foundations of the Design-Approach

A finite population of size N is a set U of N distinct elements, also called units, i.e., $U = \{u_1, \ldots, u_N\}$. A sample of size n is a subset of n elements of U. The size of a sample might be fixed ahead of sampling, or determined during the sampling, according to the values of observable variables $X(u_i)$. In the classical Design Approach the variables $X(u)$ are called the values of the units. In the modern Modeling Approach, $X(u)$ are called parameters, or quantities. Let Ω denote the set of all possible samples of a given kind (or structure). This set is called the sample-space. Since N is finite, Ω contains a finite number of different samples s. Each such sample can be assigned a probability, $P(s)$, according to the random mechanism of choosing s. For example, in a simple random sampling without replacement, of size n, n different integers are chosen at random from the set $1, 2, \ldots, N$, and the corresponding units of U are gathered into the sample s. There are $\binom{N}{n}$ different ways of choosing such samples, and therefore each element of Ω is assigned the same probability, $P(s) = 1/\binom{N}{n}$. This is the probability function for simple random sampling, without replacement, in the Design Approach. Let $Y_j = X(u_j)$, $j = 1, \ldots, N$. In addition, let $\theta = \theta(Y_1, \ldots, Y_N)$ be a parametric function of the population values. A statistic associated with a sample of size n is a list of the sample units and the values of these units, we denote such a statistic by $T = (s, \mathbf{Y}_s)$, where $s = (u_{i_1}, \ldots, u_{i_n})$, and $\mathbf{Y}_s = (Y_1, \ldots, Y_n)$. A statistic $\hat{T}_\theta = (s, \theta(\mathbf{Y}_s))$ is an unbiased estimator of θ, if

$$E_P\{\hat{T}_\theta\} = \sum_{s \in \Omega} P(s)\hat{T}_\theta = \theta. \tag{6.1}$$

For example, if $\theta(Y_1, \ldots, Y_N) = \sum_{i=1}^{N} Y_i$ and if $\hat{T}_\theta = \frac{N}{n}\mathbf{Y}'_s 1_n$, where s is a simple random sample, then $\hat{T}_\theta$ is an unbiased estimator of the population sum θ.

In a similar manner we can define the mean-squared error (MSE) of an estimator, i.e.,

$$\text{MSE}\{\hat{T}_\theta\} = \sum_{s \in \Omega} P(s)(\hat{T}_\theta - \theta)^2. \tag{6.2}$$

As in the above example, for sampling at random, without replacement,

$$MSE\{\hat{T}_\theta\} = \text{Var}\{\hat{T}_\theta\} = \frac{N^2}{n} S_y^2 \left(1 - \frac{n-1}{N-1}\right),$$

where S_y^2 is the population variance of the Y values.

The estimation under the design approach might be complicated. The reader is recommended to see the paper of Zacks (1981a) on Bayes and equivariant estimators of the variance of a finite population, when sampling is simple random. See also Cochran (1977), Des Raj (1972) and Hedayat and Sinha (1991). On the relevance of randomization in survey sampling see the paper of Basu (1978).

6.2 Foundation of the Modeling-Approach

In the modeling approach, the value of each element of the population consists of a deterministic component and a random one, i.e., $Y_i = \eta_i + E_i$, where E_i is a random variable with mean 0 and finite variance $\sigma_i^2 (i = 1, 2, \dots, N)$. This model can be written as $\mathbf{Y}_N = \boldsymbol{\eta}_N + \mathbf{E}_N$, where the bold-faced vectors are N-dimensional, and $E\{\mathbf{E}_N\} = 0$; and the variance–covariance matrix of $\mathbf{E}_N$ is an $N \times N$ symmetric, positive definite matrix $\mathbf{V}_N$.

In many cases, the elements of the population have known covariates, $\mathbf{x}_i = (x_{i1}, x_{i2}, \dots, x_{ip})'$, and the connection between η_i and $\mathbf{x}_i$ is linear, $\eta_i = \mathbf{x}_i'\boldsymbol{\beta}$. Thus, if $\mathbf{Y}_N$ is a vector of all population values, the general linear model is

$$\mathbf{Y}_N = \mathbf{X}_N'\boldsymbol{\beta} + \mathbf{E}_N, \tag{6.3}$$

where $\mathbf{X}_N$ is an $N \times p$ matrix whose rows are $\mathbf{x}_i'$, ($i = 1, \dots, N$). $\mathbf{E}_N$ is an N-dimensional random vector, with expected value zero and covariance matrix $\mathbf{V}_N = \sigma^2 \mathbf{W}_N$. The parameters of this linear model are denoted as $\psi(\boldsymbol{\beta}, \mathbf{V})$.

There are six linear models mentioned in the book by Bolfarine and Zacks (1992). It is often possible to express a population quantity θ as a sum of θ_s and θ_r, where θ_s is a function of the sample values, while θ_r is a function of the population elements which are not in the sample. The following are a few examples:

(I) Suppose that θ is the population mean value, then $\theta_s = \frac{n}{N} \bar{y}_s$, where $\bar{y}_s$ is the mean of the sample values, and $\theta_r = \left(1 - \frac{n}{N}\right) \bar{y}_r$, where $\bar{y}_r$ is the mean of the sub-population which is disjoint from the sample. Then, $\theta = \theta_s + \theta_r$.

(II) The ratio predictor of the population total, when values of a covariate $\mathbf{x}$ are known, namely

$$\hat{T}_R = n\bar{y}_s + (N - n)\bar{x}_r \frac{\bar{y}_s}{\bar{x}_s}. \tag{6.4}$$

Here, $\bar{y}_s$ is the sample mean of the Y values, $\bar{x}_r$ is the mean of x in the remainder, and $\bar{x}_s$ is the sample mean of the x values. In this example $\theta_s = n\bar{y}_s$ and $\theta_r = (N - n)\bar{x}_r \frac{\bar{y}_s}{\bar{x}_s}$. In this example θ_r is completely known after sampling. The ratio predictor (6.4) is based on the linear model $\mathbf{Y} = \beta \mathbf{X} + \mathbf{e}$, which implies that the predictor of $\bar{y}_r$ is $\bar{x}_r \frac{\bar{y}_s}{\bar{x}_s}$.

(III) The population variance

$$\theta = S_y^2 = \frac{1}{N} \sum_{i=1}^{N} (Y_i - \bar{Y}_N)^2 = \theta_s + \theta_r. \tag{6.5}$$

In this case,

$$\theta_s = \frac{n}{N} s_y^2, \tag{6.6}$$

where $s_y^2 = \frac{1}{n} \sum_{i \in s} (Y_i - \bar{Y}_s)^2$ and

$$\theta_r = \left(1 - \frac{n}{N}\right) \left[S_{ry}^2 + \frac{n}{N} (\bar{Y}_s - \bar{Y}_r)^2 \right], \tag{6.7}$$

where $S_{ry}^2 = \frac{1}{N-n} \sum_{i \in r} (y_i - \bar{y}_r)^2$.

We recommend that the reader will consider also the paper of Zacks (2002) entitled: "In the Footsteps of Basu: The Predictive Modeling Approach to Sampling from Finite Populations."

The theory outlined below is called "*prediction theory for finite populations*".

6.3 Optimal Predication of Population Quantities

6.3.1 Best Linear Unbiased Predictors

Without loss of generality, assume that the $\mathbf{Y}_N = (\mathbf{Y}_s', \mathbf{Y}_r')'$. Correspondingly, consider the partitions

$$\mathbf{X}_N = \begin{pmatrix} \mathbf{X}_s \\ \mathbf{X}_r \end{pmatrix} \text{ and } \mathbf{V} = \begin{pmatrix} \mathbf{V}_s & \mathbf{V}_{sr} \\ \mathbf{V}_{rs} & \mathbf{V}_r \end{pmatrix}.$$

The population total is $T = 1_s' \mathbf{Y}_s + 1_r' \mathbf{Y}_r$. A linear predictor of T is $\hat{T}_L = 1_s' \mathbf{Y}_s + l_{sr}' \mathbf{Y}_s$, where l_{sr} is a vector of linear coefficients. The predictor $\hat{T}_L$ is called ψ-unbiased if and only if,

$$E_\psi (T - \hat{T}_L) = 0 \text{ for } \text{ all } \psi = (\boldsymbol{\beta}, \mathbf{V}). \tag{6.8}$$

Or if and only if $1'_r \mathbf{X}_r = \mathbf{l}'_{sr} \mathbf{X}_s$.
The ψ-MSE of a linear predictor is

$$E_\psi \{ \left(T - \hat{T}_L \right)^2 \} = E_\psi \left\{ \left[\left(\mathbf{l}_{sr} - \mathbf{V}_s^{-1} \mathbf{V}_{sr} 1_r \right)' (\mathbf{y}_s - \mathbf{X}_s \boldsymbol{\beta}) \right]^2 \right\}$$

$$+ 1'_r \mathbf{V}_r 1_r - 1'_r \mathbf{V}_{rs} \mathbf{V}_s^{-1} \mathbf{V}_{sr} 1_r + \left[\left(1'_{sr} \mathbf{X}_s - 1'_r \mathbf{X}_r \right) \boldsymbol{\beta} \right]^2 . \tag{6.9}$$

Formula (6.9) can be obtained by writing

$$(T - \hat{T}_L)^2 = \left[\mathbf{l}'_{sr} \mathbf{y}_s - \mathbf{l}'_{sr} \mathbf{X}_s \boldsymbol{\beta} - 1'_r \mathbf{V}_{rs} \mathbf{V}_s^{-1} \mathbf{V}_{sr} 1_r + 1'_r \mathbf{V}_{rs} \mathbf{V}_s^{-1} \mathbf{V}_{sr} 1_r \right]^2 .$$

Furthermore, if $\hat{T}_L$ is ψ-unbiased, its ψ-MSE is

$$E_\psi \{ (T - \hat{T}_L)^2 \} = \mathbf{l}'_{sr} \mathbf{V}_s \mathbf{l}_{sr} - 2\mathbf{l}'_{sr} \mathbf{V}_{sr} 1_r + 1'_r \mathbf{V}_r 1_r . \tag{6.10}$$

The *best unbiased linear* predictor is

$$\hat{T}_{\text{BLU}} = 1'_s \mathbf{y}_s + 1'_r \left[\mathbf{X}_r \hat{\boldsymbol{\beta}}_s + \mathbf{V}_{rs} \mathbf{V}_s^{-1} (\mathbf{y}_s - \mathbf{X}_s \hat{\boldsymbol{\beta}}_s) \right], \tag{6.11}$$

where

$$\hat{\boldsymbol{\beta}}_s = (\mathbf{X}'_s \mathbf{V}_s^{-1} \mathbf{X}_s)^{-1} \mathbf{X}'_s \mathbf{V}_s^{-1} \mathbf{y}_s \tag{6.12}$$

is the weighted least squares estimator of $\boldsymbol{\beta}$, based on the sample values. The prediction variance of $\hat{T}_{\text{BLU}}$ is

$$\text{Var}_\psi \{ (\hat{T}_{\text{BLU}} - T)^2 \} = 1'_r \mathbf{V}_r 1_r - 1'_r \mathbf{V}_{rs} \mathbf{V}_s^{-1} \mathbf{V}_{sr} 1_r$$

$$+ 1'_r (\mathbf{X}_r - \mathbf{V}_{rs} \mathbf{V}_s^{-1} \mathbf{X}_s)(\mathbf{X}'_s \mathbf{V}_s^{-1} \mathbf{X}_s)^{-1} (\mathbf{X}_r - \mathbf{V}_{rs} \mathbf{V}_s^{-1} \mathbf{X}_s)' 1_r . \tag{6.13}$$

Notice that in the above formulas, the population variance–covariance matrix V is assumed to be known. For additional information see the book of Bolfarine and Zacks (1992), and the paper of Bolfarine and Zacks (1994) on the optimal prediction of the finite population regression coefficients.

6.4 Bayes Prediction in Finite Populations

6.4.1 General Theory

In the present section, the linear model (6.3) is further specified by assuming the error vector $\mathbf{E}$ is normally distributed, i.e., $\mathbf{E} \sim N(0, \mathbf{V})$ and $\boldsymbol{\beta} \sim N(\mathbf{b}, \mathbf{B})$. Moreover, we assume that $\mathbf{V} = \sigma^2 \mathbf{W}$, where $\mathbf{W}$ is known and block diagonal, i.e., $\mathbf{W}_{sr} = \mathbf{W}'_{rs} = 0_{sr}$. This model is designated as $\psi_B = \psi(\mathbf{b}, \mathbf{B}, \sigma)$.

The Bayesian estimator of $\theta(\mathbf{Y})$, for the squared error loss, is the mean of the posterior distribution,

$$\hat{\theta}_B(\mathbf{Y}_s) = E_{\psi_B}\{\theta(\mathbf{Y})|\mathbf{Y}_s\}, \tag{6.14}$$

with Bayesian prediction risk (the mean of the posterior risk)

$$E_{\psi_B}\{(\hat{\theta}_B(\mathbf{Y}_s) - \theta(\mathbf{Y}))^2\} = E_{\psi_B}\{\mathrm{Var}_{\psi_B}\{\theta(\mathbf{Y})|\mathbf{Y}_s\}\}. \tag{6.15}$$

The following results were proved in the paper of Bolfarine and Zacks (1991a).

Result 1 Under the Bayesian model ψ_B, the posterior distribution of $\mathbf{Y}_r$, given $\mathbf{Y}_s$, is multivariate normal with mean vector

$$E_{\psi_B}\{\mathbf{Y}_r|\mathbf{Y}_s\} = \mathbf{X}_r \hat{\boldsymbol{\beta}}_B + \mathbf{V}_{rs}\mathbf{V}_s^{-1}(\mathbf{Y}_s - \mathbf{X}_s\hat{\boldsymbol{\beta}}_B), \tag{6.16}$$

where

$$\hat{\boldsymbol{\beta}}_B = (\mathbf{X}'_s\mathbf{V}_s^{-1}\mathbf{X}_s + \mathbf{B}^{-1})^{-1}(\mathbf{X}'_s\mathbf{V}_s^{-1}\mathbf{Y}_s + \mathbf{B}^{-1}\mathbf{b}). \tag{6.17}$$

Result 2 The posterior covariance matrix of $\mathbf{Y}_r$ is

$$\mathrm{Var}_{\psi_B}\{\mathbf{Y}_r|\mathbf{Y}_s\} = \mathbf{V}_r - \mathbf{V}_{rs}\mathbf{V}_s^{-1}\mathbf{V}_{sr} + (\mathbf{X}_r - \mathbf{V}_{rs}\mathbf{V}_s^{-1}\mathbf{X}_s)$$
$$\times (\mathbf{X}'_s\mathbf{V}_s^{-1}\mathbf{X}_s + \mathbf{B}^{-1})^{-1}(\mathbf{X}_r - \mathbf{V}_{rs}\mathbf{V}_s^{-1}\mathbf{X}_s)'. \tag{6.18}$$

Letting $B^{-1} \to 0$, and $V_{rs} = 0$, and let σ have a non-informative prior $\xi(\sigma) \propto 1/\sigma$ then we get for $n > p + 2$ the simplified results

$$E_{\psi_B}\{\mathbf{Y}_r|\mathbf{Y}_s\} = \mathbf{X}_r \hat{\boldsymbol{\beta}}_s, \tag{6.19}$$

$$\mathrm{Var}_{\psi_B}\{\mathbf{Y}_r|\mathbf{Y}_s\} = \hat{\sigma}^2 \frac{n-p}{n-p-2}\left[\mathbf{W}_r + \mathbf{X}_r(\mathbf{X}'_s\mathbf{W}_s^{-1}\mathbf{X}_s)^{-1}\mathbf{X}'_r\right], \tag{6.20}$$

where

$$\hat{\sigma}^2 = (\mathbf{Y}_s - \mathbf{X}_s\hat{\boldsymbol{\beta}}_s)'\mathbf{W}_s^{-1}(\mathbf{Y}_s - \mathbf{X}_s\hat{\boldsymbol{\beta}}_s). \tag{6.21}$$

The Bayesian predictor of the population total, and its prediction risk, are immediately obtained from (6.19) to (6.20). See also the paper of Bolfarine and Zacks (1991a) on Bayes and minimax prediction in finite population.

6.4.2 Bayesian Prediction of the Population Variance

As in (6.5)–(6.7), the population variance is $S_y^2 = \frac{n}{N} s_y^2 + \left(1 - \frac{n}{N}\right) \left(S_{ry}^2 + \frac{n}{N} (\bar{Y}_s - \bar{Y}_r)^2\right)$.

The Bayes predictive distribution of $\bar{Y}_r$, given $\mathbf{Y}_s$, is normal with mean $\eta_r = \frac{1}{N-1} \mathbf{1}_r' \mathbf{X}_r \boldsymbol{\beta}_s$ and variance $D_r^2 = \frac{1}{(N-n)^2} \mathbf{1}_r' \boldsymbol{\Sigma}_r \mathbf{1}_r$, where $\boldsymbol{\Sigma}_r$ is the variance–covariance matrix of $\mathbf{Y}_r$, and the distribution of $\frac{n}{N} D_r^2 \chi^2[1, \lambda]$ is proportional to the noncentral chi-squared distribution, with the parameter of noncentrality $\lambda = \frac{(\eta_r - \bar{Y}_s)^2}{2D_r^2}$. This implies that

$$E_{\psi_B}\left\{\frac{n}{N} (\bar{Y}_s - \bar{Y}_r)^2 | \mathbf{Y}_s\right\} = \frac{n}{N} (D_r^2 + (\eta_r - \bar{Y}_s)^2). \tag{6.22}$$

Let $\mathbf{A}_r = \left(I_r - \frac{1}{N-n} \mathbf{1}_r \mathbf{1}_r'\right)$ and $\boldsymbol{\mu}_r = \mathbf{X}_r \hat{\boldsymbol{\beta}}_s$. We obtain

$$E_{\psi_B}\{S_{ry}^2 | \mathbf{Y}_s\} = \frac{1}{N-n} \left[tr\{\mathbf{A}_r \boldsymbol{\Sigma}_r\} + \boldsymbol{\mu}_r' \mathbf{A}_r \boldsymbol{\mu}_r\right]. \tag{6.23}$$

Then, from (6.22) and (6.23) we infer that the Bayesian predictor of S_y^2, under model ψ_B and squared-error loss, is

$$\hat{S}_{By}^2 = \frac{n}{N} s_y^2 + \left(1 - \frac{n}{N}\right) \left[\frac{1}{N-n} \left(tr\{\mathbf{A}_r \boldsymbol{\Sigma}_r\} + \boldsymbol{\mu}_r' \mathbf{A}_r \boldsymbol{\mu}_r\right) + \frac{n}{N} \left(D_r^2 + (\eta_r - \bar{Y}_s)^2\right)\right]. \tag{6.24}$$

6.4.3 Bayes Equivariant Predictors of the Variance

Bayes equivariant estimators are more complicated, and depend on the prior distribution of Y_i. Let $\mathbf{Y}_s = (Y_1, \ldots, Y_n)'$ and $\mathbf{Y}_r = (Y_{n+1}, \ldots, Y_N)'$. Let $G = \{[\alpha, \beta] : -\infty < \alpha < \infty, 0 < \beta < \infty\}$, where $[\alpha, \beta]x = \alpha + \beta x$ be a group of transformations. Recall that $[\alpha, \beta]^{-1} = \left[-\frac{\alpha}{\beta}, \frac{1}{\beta}\right]$. This is the group of real affine transformations (translations and change of scale). Given the values of the observed sample, we standardize the sample values by $\mathbf{U}_n = [\bar{y}_s, s_y]^{-1} \mathbf{Y}_n$. Based on these standardized values, S_{ry}^2 is transformed to $S_{ru}^2 = \frac{S_{ry}^2}{s_y^2}$, and $q_{rs} = (\bar{y}_s - \bar{y}_r)^2$ is transformed to $q_u = \frac{q_{rs}}{s_{\bar{y}}^2}$. Accordingly, any Bayes equivariant predictor of S_N^2 can be written as

$$S_{BE}^2 = s_y^2 \left[\frac{n}{N} + \left(1 - \frac{n}{N}\right) E_H \{S_{ru}^2 | \mathbf{U}_n\} + \frac{n}{N} \left(1 - \frac{n}{N}\right) E_H \{q_u | \mathbf{U}_n\} \right], \qquad (6.25)$$

where $H(\mathbf{U}_n)$ is a prior distribution of $\mathbf{U}_n$. In the paper of Zacks (1981b), there is a derivation of this Bayes equivariant predictor, when the prior distribution H is exponential. The result obtained for the exponential prior is

$$S_{BE}^2 = s_y^2 \left[\frac{n}{N} + \left(1 - \frac{n}{N}\right) \left[\frac{N-n-1}{N-n} + \frac{1}{Nn} \left(2 + \frac{Nn}{N-n}\right) \right] \right.$$
$$\left. \times f^2(u_n) \frac{n^2}{(n-2)(n-3)} \right], \qquad (6.26)$$

in which

$$f(\mathbf{u}_n) = \max_{2 \leq j \leq n} \{0, -u_j, \frac{1}{n-j+1} \sum_{i=2}^{j} u_i\}. \qquad (6.27)$$

Simulation estimates show that the predictor S_{BE}^2 has smaller prediction risk than the classical unbiased estimator. See also the paper of Bolfarine and Zacks (1991a) on the equivariant prediction of the population variance under location and scale super-population model. See also the papers of Bolfarine and Zacks (1991b); Bolfarine et al. (1993), on minimax and asymptotic estimation; and Zacks and Rodriguez (1986).

6.5   Bayesian Allocation of Stratified Sampling for Estimating the Population Proportion

Consider a population consisting of k strata of known sizes $N_1, \ldots, N_k$. Each stratum has an unknown number of units with special characteristic (say "defective"), denoted by $\{M_i, i = 1, \ldots, k\}$. We wish to estimate the parameter $\theta = \sum_{i=1}^{k} \lambda_i P_i$, where $P_i = \frac{M_i}{N_i}$. $\lambda' = (\lambda_1, \ldots, \lambda_k)$ is prescribed. We use a stratified sampling, in which k independent simple random samples, without replacement, are drawn. The size of each sample is n_i. The cost of sampling a unit from the i-th stratum is c_i. We wish to allocate the sample sizes so that the variance of the unbiased estimator of θ will be minimized, under the budget constraint $\sum_{i=1}^{k} c_i n_i \leq C$.

Let X_i denote the number of "defective" units in the i-th sample. The unbiased estimator of θ is $\hat{\theta} = \sum_{i=1}^{k} \frac{\lambda_i X_i}{n_i}$, having a variance

$$\text{Var}\{\hat{\theta}\} = \sum_{i=1}^{k} \lambda_i^2 (P_i(1 - P_i)/n_i)(1 - (n_i - 1)/(N_i - 1)). \qquad (6.28)$$

Using the Lagrange multiplier technique, the optimal allocation, which minimizes (6.28) under the above budget constraint, is approximately

$$n_i^0 \simeq Ca_i \frac{\sqrt{P_i(1-P_i)/c_i}}{\sum_{i=1}^k a_i \sqrt{c_i P_i(1-P_i)}},$$
(6.29)

where $a_i = |\lambda_i| \sqrt{\frac{N_i}{N_i-1}}$, $i = 1, \ldots, k$. The sample size n_i^0 is the largest integer smaller or equal to the right-hand side of (6.29). The problem is that the values of P_i are unknown. We show here a Bayesian optimal allocation, for a uniform priors for M_i.

Assume that $M_1, \ldots, M_k$ are priorly independent having uniform prior distributions with probability functions

$$\xi_i(m) = I\{m = 0, 1, 2, \ldots, N_i\} \frac{1}{N_i + 1}.$$
(6.30)

Using the identity

$$\sum_{r=0}^{N-n} \binom{r+x}{r} \binom{N-x-r}{n-x} = \binom{N+1}{N-n}.$$
(6.31)

We obtain the predictive density of X_i

$$f(x|\xi_i, N_i, n_i) = \left(\frac{1}{N_i+1}\right) \sum_{m=x}^{N_i-n_i+x} \frac{\binom{m}{x}\binom{N_i-m}{n_i-x}}{\binom{N_i}{n_i}} = \frac{1}{n_i+1}, x = 0, \ldots, n_i.$$
(6.32)

The posterior density of M_i given $\{X_i = x\}$ is

$$\xi_i(m|N_i, n_i, x) = \frac{\binom{m}{x}\binom{N_i-m}{n_i-x}}{\binom{N_i+1}{N_i-n}}, m = x, \ldots, N_i - n_i + x.$$
(6.33)

The Bayesian estimator of P_i for the squared-error loss is

$$\hat{P}_i = \frac{x+1}{n_i+2}\left(1 + \frac{2}{N_i}\right) - \frac{1}{N_i}.$$
(6.34)

Accordingly, the Bayesian estimator of $\theta = \lambda' P$ is

$$\hat{\theta}_B = \sum_{i=1}^k \lambda_i \left[\left(\frac{X_i+1}{n_i+2}\right)\left(1 + \frac{2}{N_i}\right) - \frac{1}{N_i}\right].$$
(6.35)

In the special cases of estimating the total number of "defectives," $\theta = \sum_{i=1}^{k} M_i$, the Bayes estimator is

$$\hat{\theta}_B = \sum_{i=1}^{k} \frac{N_i + 2}{n_i + 2} (X_i + 1) - k. \tag{6.36}$$

We derive now the posterior variance of θ, given $(X_1, \ldots, X_k)$. Notice that

$$\frac{1}{\binom{N+1}{N-n}} \sum_{m=x}^{N-n+x} m^2 \binom{m}{x} \binom{N-m}{n-x} = (N+2)(N+3) \frac{(x+1)(x+2)}{(n+2)^2(n+3)}$$

$$- 3(N+2) \frac{x+1}{n+2} + 1. \tag{6.37}$$

Thus, we obtain

$$\text{Var}\{\theta | n, X_1, \ldots, X_n\} = \sum_{i=1}^{k} \lambda_i^2 \frac{(X_i + 1)(n_i + 1 - X_i)(N_i + 2)}{(n_i + 2)^2(n_i + 3)N_i} \left(1 - \frac{n_i}{N_i}\right). \tag{6.38}$$

The following identity

$$\frac{1}{n+1} \sum_{i=1}^{n+1} \frac{y(n+2-y)}{(n+2)^2(n+3)} = \frac{1}{6(n+2)}$$

and (6.38) yield the prior Bayes risk

$$R(\xi, \mathbf{n}) = \frac{1}{6} \sum_{i=1}^{k} \lambda_i^2 \frac{1}{n_i + 2} \frac{N_i + 2}{N_i} \left(1 - \frac{n_i}{N_i}\right). \tag{6.39}$$

$$n_i^0 \simeq \left[\frac{\left(C + 2\sum_{i=1}^{k} c_i\right) \left(\frac{|\lambda_i|\left(1 + \frac{2}{N_i}\right)}{\sqrt{c_i}}\right)}{\sum_{i=1}^{k} |\lambda_i| \left(1 + \frac{2}{N_i}\right) \sqrt{c_i}} - 2 \right]^+. \tag{6.40}$$

For details and further proofs, see the paper of Zacks (1970a).

6.6 Bayesian Sequential Estimation of the Size of a Finite Population

The problem of estimating the size of a finite population is of interest in ecology, wild life management, census surveys, software reliability, and more. The reader is referred to the book of Seber (1985) for a discussion of the various methods developed during the years. We present here the results of Zacks, Pereira, and Leite (1990), on the estimation of the size of a finite, closed population, by the capture–recapture method. We consider a method, which is applicable in fishery, in which the size of the captured samples is also random. It is assumed that each drawn sample is a simple random one, without replacement. Each element of a sample is tagged, if it was not tagged before, and the number of newly tagged elements is recorded. Sampling is continued until a given stopping rule is satisfied. The problem of choosing a stopping rule will be discussed later.

6.6.1 The Statistical Model

Consider a population of (unknown) size N. We assume that during the sampling process the population is closed (does not change it size dynamically). Samples are drawn at random and without replacement sequentially. Let M_k be the size (possibly random) of the k-th sample. Let U_k denote the number of un-tagged elements in the k-th sample, and let $T_k = \sum_{i=1}^{k} U_i$. Obviously, $U_1 = M_1$ and $0 \leq U_k \leq M_k$. Also, $T_k \leq N$, and $M^* \leq \max_{1 \leq i \leq k}\{M_i\} \leq N$. Since the captured samples are returned to the population after tagging, we assume that the sizes of the consecutive samples are conditionally independent of the sizes of the previous ones.

Let $\mathbf{D}_k = (U_1, \ldots, U_k)$ and $\mathbf{M}_k = (M_1, \ldots, M_k)$. The likelihood function of N, given $\mathbf{D}_k$ and $\mathbf{M}_k$, is

$$L(N; \mathbf{D}_k, \mathbf{M}_k) = \prod_{j=1}^{k} \frac{\binom{N - T_{j-1}}{U_j}\binom{T_{j-1}}{M_j - U_j}}{\binom{N}{M_j}}$$

$$= \left[\frac{\prod_{j=1}^{k} \binom{T_{j-1}}{M_j - U_j}}{\prod_{j=1}^{k} U_j!} \right] I\{A_k(T_k) K(N; k, M_k, T_k)\}, \qquad (6.41)$$

where,

$$K(N; k, \mathbf{M}_k, T_k) = \frac{I\{N \geq T_k\} N!}{\left[(N - T_k)! \prod_{j=1}^{k} \binom{N}{M_j} \right]} \qquad (6.42)$$

and where $I\{A_k(x)\}$ is the indicator of the set

$$A_k(x) = \{x; x = 0, 1, \ldots, \max_{1 \le j \le k}\{M_j\} \le x \le \min\{N, \sum_{j=1}^{k} M_j\}. \tag{6.43}$$

$K(N; k, \mathbf{M}_k, T_k)$ is the likelihood kernel, and the minimal sufficient statistic for N after k samples is T_k. The p.d.f. of T_k is

$$P\{T_k = t | N = n, \mathbf{M}_k\} = K(N; k, \mathbf{M}_k, t)I\{A_k(t)\} \sum_{i=0}^{t} (-1)^{t-1} \frac{\prod_{j=1}^{k}\binom{t}{M_j}}{i!(t-i)!}. \tag{6.44}$$

6.6.2   The Bayesian Framework

Let $\pi(n)$ denote a prior p.d.f. for N on the nonnegative integers. The posterior p.d.f. of N is

$$\pi(n|k, \mathbf{M}_k, t) = \frac{\pi(n)K(n; k, \mathbf{M}_k, t)I_t(n)}{\sum_{n=t}^{\infty} \pi(n)K(n; k, \mathbf{M}_k, t)}, \tag{6.45}$$

where $I_t(n) = I\{n : n \ge t\}$. This posterior p.d.f. is not well defined if $\pi(n) = 0$, for all $n \ge t$. This will not be a problem if $\pi(n) > 0$ for all $n \ge 0$. If $M_j = 1$, for all $j \ge 1$, an improper prior like $\pi(n) = c$, for all n, will yield $\sum_{n=t}^{\infty} \pi(n)K(n; k, \mathbf{M}_k, t) = c \sum_{n=t}^{\infty} \frac{n!}{(n-t)!n^k}$ which is finite only if $t \le k - 2$. Thus, such an improper prior should not be considered. We also proved that if $\sum_{n=0}^{\infty} \pi(n) < \infty$, and if $M_k^* \le t \le \sum_{j=1}^{k} M_j$, then $\sum_{n=t}^{\infty} \pi(n)K(n; k, \mathbf{M}_k, t) < \infty$.

6.6.3   Bayesian Estimator of N

The Bayesian estimator of N for the squared-error loss is the mean of the posterior distribution, i.e.,

$$\hat{N}_B = \sum_{n=t}^{\infty} n\pi(n|k, \mathbf{M}_k, t) \equiv B(k, t). \tag{6.46}$$

If the prior distribution is concentrated on one point, n^*, then the Bayesian estimator is equal to n^*, without sampling. Let $R(k, t)$ denote the posterior variance of N, given $T_k = t$. In the following we provide a few results (theorems) which were proved in Zacks, Pereira, and Leite (1990). See also Leite and Pereira (1987).

Result 1 For all $k \geq 2$, assume that $M_k^* \leq t \leq S_k$, where $S_k = \sum_{j=1}^{k} M_j$, then

$$B(k, t) \geq B(k+1, t). \tag{6.47}$$

Equality holds when $\pi(n)$ is degenerate.

Result 2 For all $k \geq 2$, let $M_k^* \leq t \leq S_k - 1$, then

$$B(k, t) \leq B(k, t+1). \tag{6.48}$$

Result 3 For a fixed $t \geq M^*$, if $\pi(n)$ has a finite prior mean, then $\lim_{k \to \infty} B(k, t) = t$.

Result 4 The posterior risk (posterior variance) of N is

$$R(k, t) = \frac{K(k, M_k, t)}{K(k, M_k, t+1)} [B(k, t+1) - B(k, t)]. \tag{6.49}$$

6.6.4 Computing the Bayes Estimator and the Bayes Risk for Poisson Prior

In the present special case, the prior p.d.f. is Poisson with mean λ. We also assume that $M_j = M$, for all $j \geq 1$. Thus we obtain

$$B(k, t) = \frac{t + \lambda \sum_{n=0}^{\infty} \lambda^n a_n(k, t+1, M)}{\sum_{n=0}^{\infty} \lambda^n a_n(k, t, M)}, \tag{6.50}$$

where

$$a_n(k, t, M) = \frac{1}{n! \left(\prod_{i=1}^{M} (n + t + 1 - i)^k \right)}. \tag{6.51}$$

Formula (6.49) is inappropriate for numerical calculations, since the values of $a_n(k, t, M)$ decrease to zero very fast. Notice that $a_0(k, t, M) = 1/\prod_{i=1}^{M}(t+1-i)^k$. Moreover, since $T_1 = M$, then for any k, $t+1 > M$; therefore $a_0(k, t, M) > 0$, and there exists unique inverse $\left(\sum_{n=0}^{\infty} \lambda^n a_n(k, t, M)\right)^{-1} = \sum_{n=0}^{\infty} \lambda^n b_n(k, t, M)$. We obtain that the Bayesian estimator of N is,

$$B(k, t) = t + \lambda \sum_{n=0}^{\infty} \lambda^n a_n(k, t+1, M) x \sum \lambda^n b_n(k, t, M). \tag{6.52}$$

Recall that the product of two power series is the convolution

$$\sum_{n=0}^{\infty} a_n \lambda^n x \sum_{n=0}^{\infty} b_n \lambda^n = \sum_{n=0}^{\infty} \left(\sum_{k=0}^{n} a_k b_{n-k} \right) \lambda^n.$$

We specify now the coefficients $b_n(k, t, M)$. Starting with $b_0(k, t, M) = 1/a_0(k, t, M) = \prod_{i=1}^{M}(t+1-i)^k$. For $n \geq 1$,

$$b_n(k, t, M) = (-1)^n D_n(k, t, M), \tag{6.53}$$

where $D_n(k, t, M)$ is the Wronski determinant of

$$D_n(k, t, M) = \begin{pmatrix} d_1(k, t, M) & d_2(k, t, m) & \ldots & d_n(k, t, M) \\ 1 & d_1(k, t, m) & \ldots & d_{n-1}(k, t, m) \\ 0 & 1 & \ldots & d_{n-2}(k, t, m) \\ 0 & 0 & \ldots & . \\ 0 & 0 & 1 & d_1(k, t, M) \end{pmatrix} \tag{6.54}$$

with $d_j(k, t, M) = \frac{1}{j!} \prod_{i=1}^{M}(\frac{t+1-i}{t+j+1-i})^k$, $j = 1, \ldots, n$. The determinant of (6.54) can be computed recursively as follows:

$$D_0(k, t, M) = 1, \tag{6.55}$$

and

$$D_n(k, t, M) = \sum_{j=0}^{n-1} (-1)^j d_{j+1}(k, t, M) D_{n-j-1}(k, t, M), n \geq 1. \tag{6.56}$$

Finally, we obtain the Bayesian estimator of N, $B(k, t) = t + \lambda \sum_{n=0}^{\infty} \lambda^n B_n (k, t, M)$, where

$$B_n(k, t, M) = \sum_{l=0}^{M} n \frac{(-1)^l}{(n-l)!} \prod_{i=1}^{M} \left(\frac{t+1-i}{t+n-l+2-i} \right)^k D_l(k, t, m). \tag{6.57}$$

A few more algebraic manipulations yield the formula

$$B(k, t) = t + \lambda \frac{\sum_{n=0}^{\infty} \lambda^n \frac{1}{n!} \prod_{j=1}^{M} \left(\frac{t+1-j}{n+t+2-j} \right)^k}{\sum_{n=0}^{\infty} \lambda^n \frac{1}{n!} \prod_{j=1}^{M} \left(\frac{t+1-j}{n+t+1-j} \right)^k}. \tag{6.58}$$

In Table 6.1 we present the values of $B(k, t)$ and the risk $R(k, t)$, for the Poisson prior with means $\lambda = 20$, $\lambda = 50$, and $M = 1$. The population size is $N = 50$.

Table 6.1 Values of $B(k, t)$ and $R(k, t)$ for Poisson priors

		λ			
		20		50	
k	T_k	$B(k, t)$	$R(k, t)$	$B(k, t)$	$R(k, t)$
3	2	19.008	19.979	49.001	49.998
13	10	20.660	15.516	48.071	48.603
40	25	30.480	7.073	45.984	34.280
70	33	35.926	3.458	42.961	15.692
100	42	44.128	2.383	48.495	8.893
150	48	48.969	1.025	50.670	3.157
300	50	50.053	0.053	50.134	0.136

6.6.5 Stopping Rules

A sequential procedure requires a stopping rule, which indicates when to stop sampling.

Optimal stopping rule regarding cost consideration can be obtained in the following manner.

Let $\varrho(k, t)$ denote the risk associated with the optimal stopping policy. Let $c(M_k)$ denote the cost of observing the k-th sample. Let γ denote the cost penalty of observing one unit of posterior risk. $\varrho(k, t)$ satisfies the functional equation

$$\varrho(k, t) = \min\{\gamma R(k, t), c(M_{k+1}) + \sum_{u=0}^{M_{k+1}} g_{k+1}(u|k, \mathbf{M}_k, t)\varrho(k + 1, t + u)\},$$

(6.59)

where

$$g_{k+1}(u|k, \mathbf{M}_k, t) = \sum_{n=0}^{\infty} \pi(n|k, \mathbf{M}_k, t) \frac{\binom{n-1}{u}\binom{t}{M_{k+1}-u}}{\binom{n}{M_{k+1}}}.$$

(6.60)

It is optimal to stop after k samples, if

$$K_\varrho = \min\{k \geq 1 : \gamma R(k, t) = \varrho(k, t)\}.$$

(6.61)

As in dynamic programming, the function $\varrho(k, t)$ truncated at K, can be obtained by backwards induction, starting with

$$\varrho^{(0)}(k + K, t) = \gamma R(k + K, t)$$

(6.62)

and for $j = 1, \ldots, K$, compute

$$\varrho^{(j)}(k + K - j, t) = \min\{\gamma R(k + K - j, t), c(M_{k+K-j+1}) + \sum_{u=0}^{M_{k+K-j+1}} g_{k+K-j+1}$$

$$\times (u|k + K - j, \mathbf{M}_{k+K-j}, t)\varrho^{(j-1)}(k + K - j + 1, t + u)\}. \tag{6.63}$$

Freeman (1972) performed extensive numerical dynamic programming, when $M_k = 1$, and $\pi(n) \propto n^{-\nu}$, $\nu \geq 2$. The stopping boundaries obtained by Freeman are approximately linear, i.e.,

$$K_L = \min\{k \geq 1 : T_k \leq b_0 + b_1 k\}. \tag{6.64}$$

We simplified by considering the stopping rule

$$K_\epsilon = \min\{k \geq 2 : R(k, T_k) < \epsilon\}. \tag{6.65}$$

The operating characteristics of the stopping rule K_ϵ will be studied below. We present first numerical computations of the boundary values $b_m^*(k)$, computed for the Poisson prior, and various $M = m$ values (Table 6.2).

For a given $M = m$, as soon as $T_k < b_m^*(k)$ we stop sampling. It is interesting to notice that the boundaries for K_ϵ are similar to those of Freeman (1972), which were obtained by dynamic programming. Let

$$K(b_m) = \min\{k : T_k \leq b_m(k)\}. \tag{6.66}$$

Define

$$h_k^{(N,m)}(t) = P_N\{T_k = t, K(b_m) \geq k\}. \tag{6.67}$$

Table 6.2 The boundary $b_m^*(k)$ for Poisson prior with $\lambda = 50$, $m = 5, 10, 15, 20$

k	$m = 5$	$m = 10$	$m = 15$	$m = 20$
3	0	11	16	22
4	0	12	18	25
5	6	13	20	27
6	7	15	23	30
7	8	16	25	33
8	8	18	27	36
9	9	19	29	39
10	10	21	32	43
15	14	29	44	58
20	18	37	55	74
25	22	45	67	90
30	26	53	79	106
35	30	61	91	122
40	34	69	104	138
50	42	85	128	170

Set $h_1^{(N,m)}(t) = I\{t = m\}$. Since $T_k = T_{k-1} + U_k$, we obtain for $k \geq 2$,

$$h_k^{(N,m)}(t) = \sum_{u=b_m(k-1)+1}^{\min\{m(k-1),t\}} h_{k-1}^{(N,m)}(u) P_N\{U_k = t - u | T_k - 1 = u\}. \tag{6.68}$$

Moreover,

$$P_N\{U_k = t - u | T_{k-1} = u\} = \frac{\binom{N-u}{t-u}\binom{u}{m-t+u}}{\binom{N}{m}}, \tag{6.69}$$

is the hyper-geometric p.d.f. $h(j; N, A, m) = \frac{\binom{A}{j}\binom{N-A}{m-j}}{\binom{N}{m}}$, $j = 0, 1, \ldots, m$.

Since $h_1^{(N,m)}(t) = I\{t = m\}$, we get

$$h_2^{(N,m)}(t) = I\{m \leq t \leq 2m\}h(t - m; N, N - m, m), \tag{6.70}$$

and, for every $k \geq 3$, we compute recursively

$$h_k^{(N,m)}(t) = \sum_{u=b_m(k-1)+1}^{\min\{m(k-1),t\}} h_{k-1}^{(N,m)}(u)h(t - u, N, N - u, m). \tag{6.71}$$

Finally,

$$P\{K(b_m) = k\} = \sum_{j=b_m(k-1)+1}^{b_m(k)} h_k^{(N,m)}(j). \tag{6.72}$$

At stopping, if $K(b_m) = \check{k}$, $T_{\check{k}}$ can assume values in the set $\{b_m(\check{k} - 1) + 1, \ldots, b_m(\check{k})\}$. The Bayes estimator assumes the value $B(\check{k}, T_{\check{k}})$ (Table 6.3).

Table 6.3 Probability distribution of $B(\check{k}, T_{\check{k}})$ and $K(b_m)$, $N = 50$, $m = 15$, $\epsilon = 0.1$				
k	T_k	$B(\check{k}, T_{\check{k}})$	$P\{B(\check{k}, T_{\check{k}})\}$	$P\{K(b_m) = k\}$
12	43	43.068	0.000002	
	44	44.078	0.000015	
	45	45.089	0.000248	
	46	46.100	0.002899	0.003164
13	47	47.077	0.007771	
	48	48.087	0.069500	
	49	49.098	0.313078	0.390349
14	50	50.077	0.606488	0.606488
			1.000000	1.000000

Chapter 7
Visibility in Random Fields

As explained in the introduction (Sect. 1.7) the problem of visibility in random fields of obscuring elements is of interest in various areas of communication or military operations research. In the 1980s I published, with Professor M. Yadin, several papers on related subjects (see Yadin and Zacks 1982, 1985, 1988, 1994). In the present chapter the theory of stochastic visibility will be presented in a brief manner. The reader is referred to the book of Zacks (1994), in which the theory is explained with many examples, accompanying figures and with solved exercises. In the present chapter we present the ideas and techniques. We focus our attention on visibility in the plane, where obstructing elements are randomly dispersed in a region as a Poisson random field. Visibility probabilities in three-dimensional spaces are discussed in the paper of Yadin and Zacks (1982). It is an important generalization of the theory about visibility in the plane, but is not discussed in the present chapter.

7.1 Poisson Random Fields and Visibility Probabilities in the Plane

Consider a countable number of disks scattered randomly in the plane. Each disk is specified by a vector (ρ, θ, y), where (ρ, θ) are the polar coordinates of the location of the center of a disk, and y is the size of its diameter. Let Υ_0 denote the σ-field of subsets of $\{(\rho, \theta, y) : 0 \leq \rho < \infty; -\pi \leq \theta < \pi; 0 \leq y < y*\}$. Let $N(B)$ denote the number of disks, whose centers belong to the set $B \epsilon \Upsilon_0$. Υ_0 is called a *Poisson random field*, if it satisfied the following conditions:

I. $N(B)$ has a Poisson distribution with mean

$$v(B) = \mu \iiint_B dG(y|\rho, \theta)dH(\rho, \theta), \qquad (7.1)$$

© Springer Nature Switzerland AG 2020
S. Zacks, *The Career of a Research Statistician*, Statistics for Industry, Technology, and Engineering, https://doi.org/10.1007/978-3-030-39434-9_7

where $G(y|\rho,\theta)$ is the conditional distribution of the diameter of a disk, given the location of its center. $H(\rho,\theta)$ is the distribution of the locations of disks.

II. If $B_1 \cap B_2 = \varnothing$, then $N(B_1)$ and $N(B_2)$ are independent random variables, and $N(B_1 \cup B_2) = N(B_1) + N(B_2)$.

The special case where $dH(\rho,\theta) = \rho d\rho d\theta$, and $G(y|\rho,\theta) = G(y)$, is called the *standard case*. In a standard case,

$$v(B) = \mu \int_0^{y*} A_B(y)dG(y), \tag{7.2}$$

where $A_B(y)$ is the area of the region in which disks with diameter y, satisfying condition B, are centered. μ is the mean number of disks centered in a region of area 1. In shadowing processes we require that the source of light (at the origin) is not covered by a disk. For this purpose we introduce an additional requirement that $\rho > y/2$.

Let P be a point in the plane. P is *visible* if the line segment OP does not intersect any random disk. The set of all visible points in a direction s, $-\pi \le s < \pi$, starting at the origin, is called *a line of sight*, L_s. Let $P = (r, s)$, and let $B(r, s)$ be the set of all disks having distances from OP smaller than $\frac{y}{2}$. These disks intersect OP and cast shadows on P. Let $C_0 = \{(\rho,\theta,y) : \frac{y}{2} \le \rho < \infty; -\pi \le \theta < \pi; 0 < y < y*\}$ be the set of all disks which do not cover the origin.

Thus, a point $P = (r, s)$ is visible if $N(B(r, s) \cap C_0) = 0$. The probability that $P = (r, s)$ is visible in a Poisson random field is

$$Q(r, s) = \exp\{-v(B(r, s) \cap C_0)\}. \tag{7.3}$$

Example 1 The source of light beam is at the origin. The distance units are meters. The target point is $T = (200, \frac{\pi}{4})$. The random field is a standard Poisson. In order that the points 0 and T will not be covered by a disk of diameter y, we add to the area of the region between the two parallel lines on the two sides of OT, which are at distance $\frac{y}{2}$ from the line of sight OT, the area of the circle, $A\{C\} = \frac{\pi y^2}{4}$. Thus, $A_B(y) + A\{C\} = 200y + \frac{\pi y^2}{4}$. We also assume that the distribution of Y is uniform on $(0.3, 0.6)$, and $\mu = 0.01$. Accordingly, $v(B(200, s) \cap C_0) = 0.01(200 * 0.45 + 0.16494) = 0.90165$. It follows that the probability that the point T is visible from O is $Q(200, \frac{\pi}{4}) = 0.4059$ (Fig. 7.1).

Let $\|L_s\| = \sup\{r : N(B(r, s) \cap C_0) = 0\}$. It follows that

$$P\{\|L_s\| > l\} = Q(l, s), 0 \le l < \infty. \tag{7.4}$$

In the standard case, $v(B(r, s) \cap C_0) = \mu\xi r$, where ξ is the expected value of the random diameter Y. Thus, according to (7.3), the distribution of $\|L_s\|$ is exponential, with mean $\frac{1}{\mu\xi}$.

Fig. 7.1 The set $B(Y)$ of centers of disks of diameter Y, which intersect $\overline{OT}$

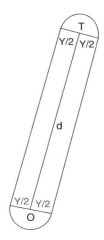

Assume that points in C_0 have coordinates (r, s, y) with $s' < s < s''$. Define

$$B_+(s) = \{(r, \theta, y) : \frac{y}{2\sin(\theta - s)} < r < \infty, s < \theta < s'', 0 < y < y*\} \qquad (7.5)$$

and

$$\dot{B}_-(s) = \{(r, \theta, y) : \frac{y}{2\sin(s - \theta)} < r < \infty, s' < \theta < s, 0 < y < y*\}. \qquad (7.6)$$

$B_+(s)$ and $B_-(s)$ are sets of all points in C_0, on the right and on the left of the ray with orientation s, of possible disks which could intersect it. We develop now formulas for the simultaneous visibility of two points.

Let P_1 and P_2 be two points with orientations $s_1 < s_2$, respectively, then the probability that these points are simultaneously visible is

$$P(s_1, s_2) = \exp\{-[v(C_0) - v((B_-(s_1) \cup B_+(s_2)) \cap C_0) \qquad (7.7)$$
$$- v((B_+(s_1) \cap B_-(s_2)) \cap C_0)]\}.$$

This formula can be generalized to the case of k points, with orientation angles $s' < s_1 < \ldots < s_k \leq s''$ as:

$$P(s_1, \ldots, s_n) = \exp\left\{ -\left[v(C_0) - v((B_-(s_1) \cup B_+(s_n)) \cap C_0) \right.\right.$$
$$\left.\left. - \sum_{i=1}^{n-1} v((B_+(s_i) \cap B_-(s_{i+1})) \cap C_0) \right] \right\}. \qquad (7.8)$$

Example 2 In order to identify a target located at a point T, one has to see a whole interval of length $L = 5$ around T. Unit length are meters. We assume that this interval is centered at T and is perpendicular to the line OT. All the points in this interval should be visible. We assume that the Poisson field is standard, and the diameter of disks is a random variable having a uniform distribution on the interval $(0.3, 0.6)$. The intensity of the field is $\mu = 0.01$. The distance between 0 and T is $d = 50$. The region, which should be clear of disks, consists of a triangle with vertices OAB where $O = (0, 0)$, $A = (\sqrt{50^2 + (2.5)^2}, \tan^{-1}(\frac{2.5}{50}))$, (the corresponding rectangular coordinates are $(2.5, 50)$), and $B = (\sqrt{2506.25}, -\tan^{-1}(\frac{1}{20}))$. On top of this triangle there is a rectangle of dimensions $5 * \frac{Y}{2}$, on each side of the lines OA and OB there are rectangles of dimensions $50.0625 * \frac{Y}{2}$, and at each one of the four corners there is a quarter of a circle of radius $\frac{Y}{2}$. The area of this region is $Area = 125 + 105.125 * \frac{Y}{2} + \frac{\pi Y^2}{4}$.

Thus $E\{Area\} = 148.8181$, and the probability that the whole interval is visible is $P = \exp(-0.01 * 148.8181) = 0.2258$ (Fig. 7.2).

Example 3 A target is detected if at least $L = 1$ m of its front is visible. A wooded area starts at $u = 50$ m from the observation point $0 = (0, 0)$. The center of the target is at point $T = (\rho, \frac{\pi}{6})$. The trees are scattered as a standard Poisson random field, with a random radius Y, which is uniformly distributed in $(0.3, 0.6)$. The intensity of the field is $\lambda = 0.001 \ 1/m^2$. How far should the target be placed within the forest, so that the detection probability would not exceed $\psi_L(s) = 0.1$?   •

In the present problem we have to find the value of ρ (the distance of T from O). The x-axis goes through O, and consider a line through T parallel to the x-axis, at distance w from it. Let $T_L T_U$ be a linear segment, perpendicular to OT. The orientation of OT_L is $s_L(\rho) = \frac{\pi}{6} - \tan^{-1}(\frac{1}{2\rho})$, and the orientation of OT_U is $s_U(\rho) = \frac{\pi}{6} + \tan^{-1}(\frac{1}{2\rho})$. As mentioned above, the beginning of the wooded area is at $u = 50$ m. The end of the wooded area is at $w(\rho) = (\rho^2 + \frac{1}{4})^{\frac{1}{2}} \cos(s_U(\rho))$. The shaded area is shown in Fig. 7.3.

Fig. 7.2 The set $B(Y)$ of centers of disks of radius Y, which obscure visibility of $T_L T_U$

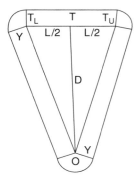

Fig. 7.3 The set $B_L(Y)$, when the field is between U and W

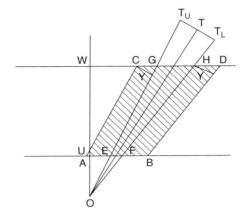

Table 7.1 Visibility probability as function of the distance ρ[m]

ρ	$\psi_L(s)$
60	0.964
80	0.679
100	0.493
120	0.362
140	0.268
160	0.200
180	0.149
200	0.117
210	0.097

The probability that the line segment $T_L T_U$ is completely visible is

$$\psi_L(s) = \exp\left\{-\lambda\frac{w^2(\rho) - u^2}{2}(\tan(s_U) - \tan(s_L))\right.$$
$$\left. -\lambda(w(\rho) - u)\frac{\cos(s_L) + \cos(s_U)}{\cos(s_L)\cos(s_U)}\mu_1\right\},$$

where $\mu_1 = E\{Y\} = 0.45$. In Table 7.1 we present $\psi_L(s)$ as a function of ρ. The answer is $\rho = 210$ m.

7.2   Visibility Probabilities on Star-Shaped Curves

A star-shaped curve is a curve in the plane that can be intersected by any ray from the origin at most once. Such a curve is specified by a continuous positive function $c(\theta)$ on the domain $[s', s'']$, where $\frac{-\pi}{2} \leq s' < s'' \leq \frac{\pi}{2}$. We also require that $c(\theta)$ will have continuous derivatives.

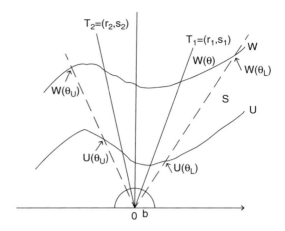

Fig. 7.4 Geometry of the field L

The field structure is specified by assuming that random disks have diameters in a closed interval $[a, b]$, where $0 \leq a < b < \infty$. We specify two star-shaped curves U and W, such that $U = \{u(\theta) : \theta' \leq \theta \leq \theta''\}$ and similarly $W = \{w(\theta) : \theta' \leq \theta \leq \theta''\}$. Moreover, for each θ, $\frac{b}{2} < u(\theta) < w(\theta) < c(\theta) - \frac{b}{2}$. We assume that all centers of disks are dispersed inside the region between U and W, which is the obstacles field. This scattering region will be denoted by C_1 (Fig. 7.4).

The expected number of disks centered in C_1 is

$$v(C_1) = \mu \int_{\theta'}^{\theta''} \int_{u(\theta)}^{w(\theta)} H(d\rho, d\theta), \tag{7.9}$$

and

$$v(C) = v(C_1) - v(B_-(s') \cap C) - v(B_+(s'') \cap C). \tag{7.10}$$

Given a point $P = (r(s), s)$ on C_1, let $\mu K_-(s, t)$ and $\mu K_+(s, t)$ denote the expected number of disks $(\rho, \theta, y) \in C_1$ with $s - t \leq \theta \leq s$ and $s \leq \theta \leq s + t$, respectively, which do not intersect the line segment OP. These functions are given by

$$K_-(s, t) = \int_{s-t}^{s} \int_{u(\theta)}^{w(\theta)} G\left(y\left(\rho, s - \theta\right) | \rho, \theta\right) H\left(d\rho, d\theta\right) \tag{7.11}$$

and

$$K_+(s, t) = \int_{s}^{s+t} \int_{u(\theta)}^{w(\theta)} G\left(y\left(\rho, s - \theta\right) | \rho, \theta\right) H\left(d\rho, d\theta\right), \tag{7.12}$$

where

$$y(\rho, \tau) = I\left\{\tau < \frac{\pi}{2}\right\} 2\rho \sin(\tau) + I\left\{\tau \geq \frac{\pi}{2}\right\}\rho. \tag{7.13}$$

Indeed, when $\|\theta - s\| < \frac{\pi}{2}$ a disk centered at (ρ, θ) does not intersect OP if its diameter is smaller than $2\rho \sin(\|\theta - s\|)$. Thus, according to (7.10) and (7.11),

$$v(C) = v(C_1) - \mu K_-(s', s' - \theta) - \mu K_+(s'', \theta'' - s''). \tag{7.14}$$

Similarly, for all $s \in [s', s'']$,

$$v(B_-(s) \cap C) = \mu K_-(s, s - \theta') - \mu K_-(s', s' - \theta') \tag{7.15}$$
$$v(B_+(s) \cap C) = \mu K_+(s, \theta'' - s) - \mu K_+(s'', \theta'' - s'').$$

Furthermore, for $s' < s_1 < s_2 < s''$,

$$v(B_+(s_1) \cap B_-(s_2) \cap C) = \mu K_+\left(s_1, \frac{(s_1 + s_2)}{2}\right) + \mu K_-\left(s_2, \frac{(s_1 + s_2)}{2}\right). \tag{7.16}$$

The value $t = (s_1 + s_2)/2$ was chosen so that a disk centered at (ρ, θ) will not intersect either OP_1 or OP_2. Define the indicator function $I(s) = 1$ if the point $(r(s), s)$ is visible, and $I(s) = 0$, otherwise. The total measure of visibility on C is

$$v(C) = \int_{s_I}^{s''} I(s)[r^2(s) + (r'(s))^2]^{\frac{1}{2}} ds. \tag{7.17}$$

The first moment of $v(C)$ is

$$E\{v(C)\} = \int_{s_I}^{s''} E\{I(s)\}[r^2(s) + (r'(s))^2]^{\frac{1}{2}} ds, \tag{7.18}$$

where

$$E\{I(s)\} = Q(r(s), s)$$
$$= \exp\{-v(C_1) + \mu[K_-(s, s - \theta') + K_+(s, \theta'' - s)]\}. \tag{7.19}$$

In the standard case

$$v(C_1) = \frac{\mu}{2} \int_{\theta'}^{\theta''} [w^2(\theta) - u^2(\theta)] d\theta. \tag{7.20}$$

For $n \geq 2$,

$$E\{V^n(C)\} = n! \int \cdots \int_{s' \leq s_1 \leq \ldots \leq s_n \leq s''} P(s_1, \ldots, s_n) \prod_{i=1}^{n} l(s_i) ds_i, \qquad (7.21)$$

where $l(s) = (r^2(s) + (r'(s))^2)^{\frac{1}{2}}$. The probability of simultaneous visibility of these n points, i.e.,

$$P(s_1, \ldots, s_n) = \exp\{-v(C_1)\} \exp \left\{ \mu K_-(s_1, s_1 - \theta') + \mu K_+(s_n, \theta'' - s_n) \right.$$

$$\left. + \mu \sum_{i=1}^{n-1} \left[K_+ \left(s_i, \frac{(s_{i+1} - s_i)}{2} \right) + K_- \left(s_{i+1}, \frac{(s_{i+1} - s_i)}{2} \right) \right] \right\}.$$

$$(7.22)$$

7.3   Some Special Cases

7.3.1   The K-Functions

In the present section we provide explicit expressions for the K-functions, under the assumptions of standard Poisson fields, and uniform distribution of Y on (a, b). Under these assumptions we obtain that $K_+(s, d\theta, v) = \Lambda(\theta - s, v, a, b)d\theta$, where

$$\Lambda(\tau, v, a, b) = I \left\{ \sin^{-1} \left(\frac{a}{2v} \right) \leq \tau < \sin^{-1} \left(\frac{b}{2v} \right) \right\}$$

$$\times \left[\frac{1}{b-a} \left(\frac{2}{3} v^3 \sin(\tau) - \frac{a}{2} v^2 + \frac{1}{24} \frac{a^3}{\sin^2(\tau)} \right) \right]$$

$$+ I \left\{ \sin^{-1} \left(\frac{b}{2v} \right) \leq \tau < \frac{\pi}{2} \right\} \left[\frac{1}{2} \left(\frac{v^2 - (a^2 + ab + b^2)}{12 \sin^2(\tau)} \right) \right]$$

$$+ I \left\{ \frac{\pi}{2} \leq \tau \right\} \left[\frac{1}{2} \left(\frac{v^2 - (a^2 + ab + b^2}{12} \right) \right].$$

$$(7.23)$$

7.3.2   The Annular Region

We simplify by assuming that $U(\theta) = u$ and $W(\theta) = w$, for all $\theta' < \theta < \theta''$, and that $b/2 < u < w$. Moreover, $K_+(s, t) = K_-(s, t) = K^*(t, w) - K^*(t, u)$, where

$$K^*(t, v) = \int_0^t \Lambda(\tau, v, a, b) d\tau$$

$$= I\left\{\sin^{-1}\left(\frac{a}{2v}\right) \leq \tau < \sin^{-1}\left(\frac{b}{2v}\right)\right\} K_1(t, v)$$

$$+ I\left\{\sin^{-1}\left(\frac{b}{2v}\right) \leq \tau < \frac{\pi}{2}\right\} K_2(t, v) \qquad (7.24)$$

$$+ I\left\{\frac{\pi}{2} \leq \tau\right\} K_3(t, v),$$

in which

$$K_1(t, v) = \frac{1}{b-a}\left[v^3\left((4v^2 - a^2)^{\frac{1}{2}} - 2v\cos(t)\right) - \frac{a}{2}v^2\left(t - \sin^{-1}\left(\frac{a}{2}\right)\right)\right.$$

$$\left. + \frac{a^2}{24}\left((4v^2 - a^2)^{\frac{1}{2}} - a\cot(t)\right)\right], \qquad (7.25)$$

$$K_2(t, v) = K_1\left(\sin^{-1}\left(\frac{b}{2v}\right), v\right) + \frac{v^2}{2}\left(t - \sin^{-1}\left(\frac{b}{1v}\right)\right)$$

$$- \frac{1}{24}(a^2 + ab + b^2)\left(\frac{(4v^2 - b^2)^{\frac{1}{2}}}{b - \cot(t)}\right), \qquad (7.26)$$

$$K_3(t, v) = K_2\left(\frac{\pi}{2}, v\right) + \left(t - \frac{\pi}{2}\right)\left(\frac{v^2}{2} - \frac{a^2 + ab + b^2}{24}\right). \qquad (7.27)$$

7.3.3 The Moments

Here we consider the target curve to be circular, over (s', s''). That is, $r(s) = r$ on C. In this case, $l(s) = r$, for all s in (s', s'').

Let $\lambda = \exp\{\frac{-\mu(w^2 - u^2)(\theta'' - \theta')}{2}\}$, and $\Psi_0(s) = \exp\{\mu[K^*(s - \theta', w) - K^*(s - \theta', u)]\}$ and $H(s) = \exp\{\mu[K^*(\theta'' - s, w) - K^*(\theta'' - s, u)]\}$.

Define recursively, for every $j \geq 1$,

$$\Psi_j(s) = \int_{s'}^s (y) \exp\left\{2\mu\left[K^*\left(\frac{s-y}{2}, w\right) - K^*\left(\frac{s-y}{2}, u\right)\right]\right\} dy. \qquad (7.28)$$

The n-th moment of $V(C)$ is then

$$\mu_n = \lambda n! \int_{s'}^{s''} \Psi_{n-1}(s) H(s) ds. \qquad (7.29)$$

The distribution of $\frac{V^*(C)}{r(s''-s')}$ is approximated by a mixture of a beta distribution with a discrete distribution centered on $(0, 1)$. This distribution is

$$F^* = I\{0 \leq x < 1\}\left[P_0 + (1 - P_0 - P_1)\left(\frac{1}{B(\alpha, \beta)}\right)\int_0^x y^{\alpha-1}(1 - y)^{\beta-1}dy\right]$$

$$+ I\{x > 1\}.$$

(7.30)

P_1 is the visibility probability

$$P_1 = \lambda \exp\{\mu \left[K^*(s' - \theta', w) - K^*(s' - \theta', u)\right.$$

$$\left. +K^*(\theta'' - s'', w) - K^*(\theta'' - s'', u)\right]\}.$$

Since three parameters are unknown, namely P_0, α, and β, we equate the first three moments $\mu_i, i = 1, 2, 3$ of $\frac{V^*(C)}{r(s''-s')}$ to those of F^*. If $\mu_n^* = \frac{\mu_n}{r^n(s''-s')^n}$ and $c_n = X$, then

$$\alpha = \frac{2c_2^2 - c_3(c_1 + c_2)}{c_1 c_3 - c_2^2},$$

$$\beta = \frac{(c_1 - c_2)(c_2 - c_3)}{c_1 c_3 - c_2^2},$$

(7.31)

$$P_0 = 1 - P_1\frac{c_1(\alpha + \beta)}{\alpha}.$$

In Table 7.2 we present the normalized moments of $V(C)$ and their beta-mixture approximation.

For each μ in this table, the lower line is the Beta-mixture approximation.

Table 7.2 Normalized moments of $V(C)$ and their beta-mixture approximation, $s' = \frac{-\pi}{18}$, $s'' = -s'$, $r = 100$ m, $u = 50$ m, $w = 75$ m, $a = 10$ m, $b = 30$ m

μ	$n = 1$	$n = 2$	$n = 3$	$n = 4$	$n = 5$	$n = 6$	∞
1	0.951	0.938	0.931	0.928	0.927	0.926	0.901
	0.951	0.938	0.931	0.927	0.924	0.922	
3	0.860	0.833	0.816	0.808	0.804	0.802	0.730
	0.860	0.833	0.816	0.803	0.794	0.787	
5	0.778	0.731	0.706	0.694	0.687	0.684	0.592
	0.778	0.731	0.706	0.691	0.680	0.672	

7.4 Random Coverage of a Circle and Distribution of Length of Visible Arcs

The distribution of the length of shadows, or the length of visible sections, on a general star-shaped curve is complicated. In the book by Zacks (1994) one can find the theory and examples when the target curve is linear. In this case, the length of a shadows depends on the location of their starting point. Shadows on the circle C are shown by black arcs. The length of shadows, or of visible arcs, is i.i.d. random variables only when the target curve is a circle around the observation point. In the present section we present this case, which was published by Yadin and Zacks (1982). See also Zacks and Yadin (1984).

7.4.1 Coverage of a Circle by Random Black Arcs

Consider a circle C of radius 1 centered at the origin. Let $A_1, \ldots, A_N$ be black arcs placed at random on C. Assume that N is a random variable, having a Poisson distribution with mean $2\pi\lambda$. We assume also that the lengths of these arcs are independent random variables $X_i, i = 1, \ldots, n$, having a c.d.f. $F(x)$ on $(0, \pi)$. Let P_τ be a point on the circle C with polar coordinates $(1, \tau), 0 \leq \tau < 2\pi$. Let $M(\tau)$ denote the number of random black arcs, covering the point P_τ. The probability that a random black arc covers a point on C is $\frac{E\{X\}}{2\pi}$. Accordingly,

$$P\{M(\tau) = j | N = n\} = b\left(j; n, \frac{E\{X\}}{2\pi}\right), \tag{7.32}$$

where $b(j; n, p)$ is the Binomial p.d.f. Thus, since N has a Poisson distribution with mean $2\pi\lambda$, we obtain that the p.d.f. of $M(\tau)$ is that of a Poisson distribution, i.e.,

$$P\{M(\tau = j\} = p(j, \lambda E\{X\}). \tag{7.33}$$

Notice that (7.31) and (7.32) are independent of the orientation τ of the point, due to the symmetry on the circle.

Let $q(t)$ denote the probability that a fixed (white) arc of length t is completely uncovered by black random arcs (completely visible). Let $q_1(t)$ denote the probability that a specified arc of length t is completely uncovered by one black random arc, given $N = 1$. Since the right end point $Y = y$ of a random black arc is uniformly distributed on C, the conditional probability that a specified arc is uncovered is, $P\{X < 2\pi - y\} = F(2\pi - y)$

$$q_1(t) = \frac{\int_t^{2\pi} F(2\pi - y) dy}{2\pi} = 1 - \frac{t + E(X) - \phi(2\pi - t)}{2\pi}, \tag{7.34}$$

where

$$\phi(t) = \int_t^{2\pi} [1 - F(x)]dx. \tag{7.35}$$

Since for $N = n$, the n random arcs are conditionally independent. Hence $q_n(t) = (q_1(t))^n$, and therefore

$$q(t) = \xi \exp\{-\lambda(t - \phi(2\pi - t))\}, \text{ for all } 0 \le t \le 2\pi, \tag{7.36}$$

where

$$\xi = \exp\{-\lambda E\{X\}\}. \tag{7.37}$$

We consider now the probability that r specified points on the circle C are not covered by black arcs. The r specified points have orientation coordinates $0 < s_1 < s_2 < \ldots < s_r < 2\pi$. Let $t_0 = 2\pi - s_r + s_1, t_i = s_i - s_{i-1}, i = 1, \ldots, r-1$. Let $N = n$, and $p_n(t_1, \ldots, t_{r-1})$ be the conditional probability that the r specified points are uncovered by n random black arcs. Obviously, $p_0(t_1, \ldots, t_{r-1}) = 1$. Furthermore, due to conditional independence, $p_n(t_1, \ldots, t_{r-1}) = (p_1(t_1, \ldots, t_{r-1}))^n$, where

$$p_1(t_1, \ldots, t_{r-1}) = \sum_{i=0}^{r-1} P\{B_i\} = 1 - \frac{r E\{X\} - \sum_{i=0}^{r=1} \phi(t_i)}{2\pi}. \tag{7.38}$$

Thus, for all $r \ge 2$, $\sum_{i=0}^{r-1} t_i < 2\pi$,

$$p(t_1, \ldots, t_{r-1}) = \xi^r \exp\left\{\lambda \sum_{i=0}^{r-1} \phi(t_i)\right\}. \tag{7.39}$$

We conclude this subsection by considering the distribution of the length of uncovered (white) arcs. Let H_s denote the length of an uncovered arc, starting at an uncovered point P_s, i.e.,

$$H_s = \sup\left\{t : \sup_{s \le \tau \le s+t} M(\tau) = 0, 0 \le t \le 2\pi\right\}. \tag{7.40}$$

We obtain,

$$P\{H_s > t | M(s) = 0\} = I\{0 \le t \le \pi\} \exp\{-\lambda t\}$$
$$+ I\{\pi < t < 2\pi\} \exp\{-\lambda t + \lambda\phi(2\pi - t)\}. \tag{7.41}$$

7.4.2   The Measure of Vacancy and Its Moments

Let P_s and P_{s+t} be two points on C, where $t > 0$. Consider the specified arc between these two points. The *measure of vacancy* on this arc is

$$Y(s, t) = \int_s^{s+t} I(\tau)d\tau, \tag{7.42}$$

where the indicator function $I(\tau) = 1$ if the point P_τ is not covered by a random black arc. Due to the symmetry on the circle, the distribution of $Y(s, t)$ does not depend on s. The expected value of $Y(s, t)$ is

$$E\{Y(s, t)\} = \int_s^{s+t} E\{I(\tau)\}d\tau = \xi t. \tag{7.43}$$

The r-th moment of $Y(s, t)$, for $r \geq 2$, is

$$\xi_r = E\{Y(s, t)\}$$

$$= \xi' r! \int_0^t d\tau (1 - \tau) \exp\{\lambda\phi(2\pi - \tau)\} \int_0^t \tau dt_1 \exp\{\lambda\phi(\tau_1)\}$$

$$\int_0^{\tau - t_1} dt_2 \exp\{\lambda\phi(\tau_2)\} \int_0^{\tau - t_1 - t_2} \cdots \tag{7.44}$$

$$\int_0^{\tau - \sum_{i=1}^{r-1} t_i} dt_{r-2} \exp\left\{\lambda\phi(t_{r-2}) + \lambda\phi\left(\tau - \sum_{i=1}^{r-2} t_i\right)\right\}.$$

The evaluation of these moments can be done recursively as follows

$$\psi_1(t) = I\{t > 0\} \exp\{\lambda\phi(t)\}. \tag{7.45}$$

For every $r \geq 2$,

$$\psi_r(t) = \int_0^t \psi_1(\tau)\psi_{r-1}(t - \tau)d\tau. \tag{7.46}$$

According to (7.43) the r-th moment is

$$\xi_r = r!\xi^r \int_0^t (t - \tau)\psi_1(2\pi - \tau)\psi_{r-1}(\tau)d\tau. \tag{7.47}$$

The r-th moment is of order of magnitude $0(t_r/r!)$. Thus, all moments of $Y(s, t)$ exist. The moment generating function (m.g.f.) satisfies the integral equation

$$\xi(v, t) = \gamma(v, t) + \xi v \int_0^t \psi_1(u)\xi(v, t - u)du, \tag{7.48}$$

where

$$\gamma(v, t) = 1 + \xi v \left(t - \int_0^t \psi_1(u)du \right). \tag{7.49}$$

Let $h(v, t)$ denote the solution of the integral equation (7.47), The Laplace transform of this solution is

$$h^*(v, w) = \frac{1}{w} + \frac{\xi v}{w^2(1 - \xi v\psi^*(w)}, \tag{7.50}$$

where $\Psi^*(w)$ is the Laplace transform of $\psi_1(t)$.

7.5   Applications to Shadowing Problem

In the present section we develop the previous results to the case where the black arcs are shadows cast by random disks in a Poisson field which is scattered inside the circle C. Moreover it is assumed that the random disks do not intersect C or cover the center O. In other words, we consider only disks with random parameter vectors in the set

$$S = \left\{ (\rho, \theta, y) : \frac{y}{2} < \rho < 1 - \frac{y}{2}; 0 < \theta < 2\pi; 0 < y < 1 \right\}. \tag{7.51}$$

Let the Poisson field be a standard one, and the c.d.f. of the diameters Y is $G(y)$. Thus, the number of disks in S has a Poisson distribution with mean

$$E\{N(S)\} = \mu \int_0^1 \int_0^{2\pi} \int_{y/2}^{1-y/2} \rho d\rho d\theta dG(y) = 2\pi\lambda, \tag{7.52}$$

where

$$\lambda = \frac{\mu}{2} \int_0^1 (1 - y)dG(y). \tag{7.53}$$

Let $X(\rho, \theta, y)$ denote the length of a shadow arc projected by a disk with parameters (ρ, θ, y). This length is given by

$$X(\rho, \theta, y) = 2 \sin^{-1} \left(\frac{y}{2\rho} \right). \tag{7.54}$$

Notice that the random length of a shadow arc does not depend on the orientation θ, only on the length of the radius, relative to ρ, the distance of its center from the origin. The maximal diameter that can yield a shadow arc of length x is

$$D(x) = \frac{2\sin\left(\frac{x}{2}\right)}{1 + \sin\left(\frac{x}{2}\right)} = 1 - \tan^2\left(\frac{\pi - x}{4}\right). \tag{7.55}$$

Thus, the c.d.f. of the shadow-arc length is

$$F(x) = \frac{\mu}{2\lambda} \int_0^{D(x)} \left(\int_{y/2\sin(x/2)}^{1-y/2} \rho d\rho \right) g(y) dy \tag{7.56}$$

$$= \frac{\mu}{2\lambda} \int_0^{D(x)} \left(1 - y - \left(\frac{y^2}{4}\right) \cot^2\left(\frac{x}{2}\right) \right) g(y) dy.$$

Consider the distribution function, for $0 \le x < \pi$

$$B_l(x) = \frac{1}{c_l} \int_0^{D(x)} y^l (1 - y)^{1 - l/2} g(y) dy, \tag{7.57}$$

where

$$c_l = \int_0^1 y^l (1 - y)^{1 - l/2} g(y) dy.$$

We can then write

$$F(x) = I\{0 < x < \pi\} \left[B_0(x) - \frac{c_2^2 B_2(x) \cot^2\left(\frac{x}{2}\right)}{4c_0} \right] + I\{x \ge \pi\}. \tag{7.58}$$

The function $\phi(t)$ defined in (7.38), when $G(y)$ is uniform on $(0, 1)$, can be written as

$$\phi(t) = -\frac{1}{3}(\pi - t) + \frac{4}{3} \tan\left(\frac{\pi - t}{4}\right) + \frac{4}{9} \tan^3\left(\frac{\pi - t}{4}\right). \tag{7.59}$$

Thus, when the Poisson field is standard, and $G(y)$ is uniform, the expected length of a shadow arc is $E\{X\} = \phi(0) = 0.7306$, and the first moment of vacancy of arc of length t is $\xi_1(t) = t \exp\{-0.7306\lambda\}$.

Chapter 8
Sequential Testing and Estimation for Reliability Analysis

8.1 Introductory Comments

Reliability of a system is defined as the probability that the system can fulfill all its functions during a specified time period without failure. The reliability of a system, as a function of time, $R(t)$, is the probability that the system functions without failure for at least t time units. In reliability theory one specifies the life length distribution of the system and its hazard function $h(t) = \frac{f(t)}{1-F(t)}$, where F and f are the c.d.f. and p.d.f. of the life distribution. In many studies the life distribution belongs to the Weibull family. The exponential distribution is a special case of Weibull, in which the hazard function is a constant. Other Weibull distributions have either monotone increasing or monotone decreasing hazard rate function. There are however many other types of life distributions. We also distinguish between repairable systems and non-repairable ones. For an introductory textbook on reliability analysis, see Zacks (1992).

In the present chapter we present results from several papers. We start in Sect. 8.2 with the Wald *sequential probability ratio test* (SPRT), which is a most efficient test procedure constrained by the probabilities of errors of type I and type II. We start with the SPRT for testing the intensity of a Poisson process. This research was motivated by consulting to an electronic industry, where failures of their product followed a homogeneous Poisson process. We proceed with two types of truncated SPRT, where truncation is according to the frequency of failures. In Sect. 8.3 we present sequential estimation of the *mean time to failure* (MTTF) in a homogeneous Poisson failure process, when the objective is to obtain estimates having specified precision. We show how to calculate the exact distribution of the stopping time and its related functionals. In Sect. 8.4 we discuss the characteristics of a *degradation process*, where a system is subjected to shocks of random strength, occurring at random times following a nonhomogeneous Poisson process. Such a system fails when the total damage exceeds a given threshold.

© Springer Nature Switzerland AG 2020
S. Zacks, *The Career of a Research Statistician*, Statistics for Industry,
Technology, and Engineering, https://doi.org/10.1007/978-3-030-39434-9_8

8.2 Sequential Testing Procedures

Sequential procedures are efficient procedures of sampling when the objective is to obtain test results or estimates with a prescribed precision. For example, in sampling inspection for the quality of a large lot of items. Generally the lot is considered acceptable if the proportion defectives (nonstandard) items in the lot are smaller than a required value p_0. On the other hand, if the proportion defectives in the lot is greater than p_1, where $p_0 < p_1$, the lot is considered non-acceptable. The problem is that the correct value of the proportion defectives is unknown, if the lot is not completely inspected. The Wald Sequential Probability Ratio Test (SPRT) is a most efficient procedure of testing the two hypotheses $H_0 : p \leq p_0$ against $H_1 : p > p_1$, when the test procedure should satisfy the two requirements: The error of type I (probability of rejecting H_0 when it is true) should be smaller than a prescribed value α; on the other hand, the error of type II (probability of accepting H_0 when H_1 is true) should be smaller than a prescribed value β.

In the electronic industry, it is often the case that the time till failure of a component has an exponential distribution with mean $\beta = \frac{1}{\lambda}$. Thus, if n identical components are put on test, and failures of components are independent, the time till the first failure is exponentially distributed, with mean $\beta_n = 1/n\lambda$. This is based on the well-known result that in a sample of n i.i.d. exponentially distributed random variables, the distribution of the first order statistic (minimum) is exponential with mean $\frac{1}{\lambda n}$. Furthermore, due to the unique property of the exponential distribution, the excess life of each of the distributions of components which have not failed is still the same exponential with parameter λ. In this trial, each component that fails is immediately replaced by another identical component. Thus, the number of observed failures in a unit time is a random variable, having a Poisson distribution, with mean $\mu = \lambda n$. Accordingly, the process of counting the number of first failures is a Poisson process with intensity $\mu = \lambda n$.

8.2.1 Wald SPRT for Poisson Processes

We start with a constructive definition of a Poisson process. Let $\{N(t), t > 0\}$ denote a counting process, which is a step function, starting at $N(0) = 0$, and increasing by a one unit jump at each epoch of failure. Such a process is called a Poisson Process with intensity λ, if the following properties hold:

  I. The process has independent increments, i.e., for any $0 \leq t_1 < t_2 < \infty$, $N(t_1)$ and $N(t_2) - N(t_1)$ are independent.
 II. For any $0 \leq t_1 < \infty$, $N(t_2) - N(t_1)$ is distributed like $N(t_2 - t_1)$.
III. For any $0 < t < \infty$, $N(t)$ has a Poisson distribution with mean λt.

The regular Poisson process is time homogeneous (property 2). Furthermore, if $\tau_1, \tau_2, \ldots$ are the jump epochs of the process, the inter-arrival times $T_n = \tau_n - \tau_{n-1}, n \geq 1$ are independent random variables, having an exponential distribution with mean $\beta = \frac{1}{\lambda}$.

For the acceptance of a product, the customer has to test if λ is sufficiently small, according to the agreement with the producer. Accordingly, a Wald SPRT test procedure is performed with $H_0 : \lambda \leq \lambda_0$ versus $H_1 : \lambda \geq \lambda_1$, and some agreed upon values of $\lambda_0 < \lambda_1$, and probabilities of errors of type I and type II are ϵ_1 and ϵ_2. The stopping boundaries are two parallel lines $B_L(t) = A_L + Bt$, and $B_U(t) = A_U + Bt$, with

$$A_U = \frac{\log\left(\frac{1-\epsilon_2}{\epsilon_1}\right)}{\log\left(\frac{\lambda_1}{\lambda_0}\right)} \tag{8.1}$$

$$A_L = \frac{\log\left(\frac{\epsilon_2}{1-\epsilon_1}\right)}{\log\left(\frac{\lambda_1}{\lambda_0}\right)}$$

$$B = \frac{\lambda_1 - \lambda_0}{\log\left(\frac{\lambda_1}{\lambda_0}\right)}.$$

The stopping time of this sequential procedure is

$$T_s = \inf\{t > 0 : (N(t) < A_L + Bt) \cup (N(t) > A_U + Bt)\}, \tag{8.2}$$

if $N(T_s) < A_L + BT_s$. then H_0 is accepted, otherwise, H_0 is rejected. Since there are two boundaries, $P\{T_s < \infty\} = 1$. The function

$$\pi(\lambda) = P\{N(T_s) < A_L + BT_s\} \tag{8.3}$$

is called the *operating characteristic function*, and is sometimes denoted by $OC(\lambda)$. Another important characteristic of the test is the expected length of the test, i.e., $E_\lambda\{T_s\}$.

In order to simplify the formulas we make the transformations $t^* = Bt$ and $\mu = \frac{\lambda}{B}$. Let $N^*(t)$ be a Poisson process, with intensity μ. Then, the problem of $N^*(t)$ crossing the boundaries $B_L^*(t^*) = A_L + t^*$ and $B_U^*(t^*) = A_U + t^*$ is similar to the original problem. The stopping times are related as $T_s = \frac{T_s^*}{B}$. Hence, without loss of generality we proceed with the parameters μ and $B = 1$. The following exact calculations of the $OC(\lambda)$ and $E\{T\}$, for truncated SPRT, were published by De and Zacks (2015).

8.2.2   *Truncated SPRT of Type I*

Notice that the parameter A_L is negative whenever $\varepsilon_1 + \varepsilon_2 < 1$. Thus, we will consider here a lower boundary $B_L(t) = -k_1 + t$ and an upper boundary $B_U(t) = k_2$, where k_1, k_2 are positive integers. Thus, define the stopping times

$$T_L^{(1)} = \inf\{t > 0 : N(t) = B_L(t)\}$$

$$T_U^{(1)} = \inf\{t > 0 : N(t) \geq k_2\} \qquad (8.4)$$

$$T_S^{(1)} = \min\{T_L^{(1)}, T_U^{(1)}\}.$$

The lower boundary intersects the upper boundary at $t = k^* = k_1 + k_2$. Moreover, $P_\lambda\{T_S^{(1)} \leq k^* - 1\} = 1$.

The stopping time $T_L^{(1)}$ is a discrete random variable with positive p.d.f. on the set of points $\{t_m = k_1 + m, m = 0, 1, \ldots, k_2\}$. Since $N(t)$ is a nondecreasing process, if $N(t)$ crosses the upper boundary first, it cannot return and cross the lower boundary. Thus, if the lower boundary is crossed first, the null hypothesis H_0 is accepted. Let $f_L(m; k_1, k_2) = P_\lambda\{T_L^{(1)} = t_m\}$ be the p.d.f. of $T_L^{(1)}$. This function can be determined recursively according to the equations

$$f_L(0; k_1, k_2) = \exp\{-\lambda k_1\}, \qquad (8.5)$$

$$f_L(m; k_1, k_2) = p(m; \lambda t_m)$$

$$- \sum_{j=0}^{m-1} f_L(j; k_1, k_2) p(m - j; \lambda(m - j)), m = 0, \ldots, k_2,$$

where $p(j; \eta) = \frac{\exp\{-\eta\}\eta^j}{j!}$ is the p.d.f. of the Poisson distribution with mean η. An analytic expression for $f_L(m; k_1, k_2)$ can be derived (see Zacks 2017); however, numerical values can be obtained very fast by using these recursive equations.

The operating characteristic function of this sequential test is

$$\pi(\lambda) = \sum_{m=0}^{k_2} f_L(m; k_1, k_2). \qquad (8.6)$$

In Table 8.1 we present some of these OC functions.

We see in this table that the OC function can be helpful in deciding the appropriate values of k_1, k_2. For example, if we wish to test whether $H_0 : \lambda \leq 0.5$ versus $H_1 : \lambda \geq 1.5$, the test with $k_1 = 3$ and $k_2 = 10$ yields error probabilities $\alpha = 0.0413$ and $\beta = 0.0631$. These values of the k parameters might be good enough.

Table 8.1 Operating characteristics, $\pi(\lambda)$, as functions of $k_1 = 3$

λ	$k_2 = 7$	$k_2 = 10$	$k_2 = 15$	$k_2 = 20$
0.1	1.0000	1.0000	1.0000	1.0000
0.3	0.9922	0.9987	0.9999	1.0000
0.5	0.9056	0.9587	0.9889	0.9968
0.7	0.6917	0.7896	0.8815	0.9297
0.9	0.4394	0.5241	0.6213	0.6872
1.1	0.2416	0.2884	0.3422	0.3790
1.3	0.1256	0.1397	0.1591	0.1702
1.5	0.0568	0.0631	0.0683	0.0705
1.7	0.0261	0.0279	0.0290	0.0293
1.9	0.0119	0.0124	0.0126	0.0126

8.2.2.1   The Distribution of $T_s^{(1)}$

The survival function $S_\lambda^{(1)}(t; k_1, k_2) = P_\lambda(T_S^{(1)} > t)$. Notice that $S_\lambda^{(1)}(t_{k_2-1}; k_1, k_2) = 0$. The survival function can be obtained in the following way. Define

$$g_S^{(1)}(j, t; k_1, k_2) = I\{(t - k_1)^+ \leq j \leq k_2 - 1\} P_\lambda\{N(t) = j, T_S^{(1)} > t\}. \qquad (8.7)$$

Then, for every $l = 1, 2, \ldots, k_2$, we have

$$g_S^{(0)}(j, t_1; k_1, k_2) = p(j, \lambda t_l) - \sum_{i=0}^{l-1} f_L(i; k_1, k_2) p(j - i, \lambda(l - i)). \qquad (8.8)$$

Notice that, $g_S^{(1)}(j, t; k_1, k_2) = I\{t \leq k_1, j \leq k_2\} p(j, \lambda t)$. On the other hand, for $t_l < t \leq t_{l+1}$,

$$g_S^{(1)}(j, t; k_1, k_2) = \sum_{i=l+1}^{j} g_S^{(1)}(i, t_l; k_1, k_2) p(j - i, \lambda(t - t_l)). \qquad (8.9)$$

Finally, since

$$S_\lambda^{(1)}(t; k_1, k_2) = I\{t \leq k_1\} P(k_2 - 1, \lambda t)$$
$$+ \sum_{l=0}^{k_2-2} I\{t_l < t \leq t_{l+1}\} \sum_{j=l+1}^{k_2-1} g_S^{(1)}(j, t; k_1, k_2), \qquad (8.10)$$

we obtain, for $l = 1, \ldots, k_2 - 2$,

$$S_\lambda^{(1)}(t_l; k_1, k_2) = P(k_2 - 1, \lambda t_l) - P(l, \lambda t_l)$$
$$- \sum_{j=l+1}^{k_2-1} \sum_{i=0}^{l-1} f_L(i; k_1, k_2) p(j - i, \lambda(l - i)). \qquad (8.11)$$

Table 8.2 Survival
probabilities

t	l	$S^{(1)}$
1		0.9999
2		0.9955
3	0	0.9665
4	1	0.7832
5	2	0.6067
6	3	0.4104
7	4	0.2272
8	5	0.0853

Here $P(j; \eta)$ is the c.d.f. of Poisson with mean η. Notice that $S_\lambda^{(1)}(t_{k_2-1}; k_1, k_2) = 0$.

In Table 8.2, we present the survival probabilities in the case of $k_1 = 3$, $k_2 = 7$, and $\lambda = 1$

The distribution function c.d.f. of $T_S^{(1)}$ is a mixture of a discrete distribution, the jumps at the epochs of t_l, and an absolutely continuous distribution, at the times of crossing the upper boundary. The density of the absolutely continuous part is defined on the intervals (t_l, t_{l+1}), $l = 0, k_2 - 2$, and is given by

$$f_U(t; k_1, k_2) = -\frac{d}{dt}\Big[I\{t \le k_1\}P(k_2 - 1, \lambda t)$$

$$+ \sum_{l=0}^{k_2-2} I\{t_1 < t < t_{l+1}S_\lambda^{(1)}(t; k_1, k_2)\Big]$$

$$= \lambda p(k_2 - 1, \lambda t) + \lambda \sum_{l=0}^{k_2-2} I\{t_l < t < t_{l+1}\} \tag{8.12}$$

$$\times \sum_{j=l+1}^{k_2-1} \sum_{i=l+1}^{j} g_S^{(1)}(j, t_l; k_1, k_2)$$

$$[p(j - i, \lambda(t - t_l)) - p(j - i - 1, \lambda(t - t_l))].$$

The expected stopping time is

$$E_\lambda\{T_S^{(1)}; k_1, k_2\} = \int_0^{k_1+k_2-1} S_\lambda^{(1)}(t; k_1, k_2)dt \tag{8.13}$$

$$= \frac{1}{\lambda} \sum_{i=0}^{k_2-1} (1 - P(i, \lambda k_1)) +$$

$$= \frac{1}{\lambda} \sum_{i=0}^{k_2-1} \sum_{j=l+1}^{k_2-1} g_S^{(1)}(j, t_l; k_1, k_2) \sum_{i=0}^{k_2-1-j} (1 - P(i, \lambda)).$$

Table 8.3 Expected stopping time $k_1 = 3$

λ	$k2 = 7$	$k2 = 10$	$k2 = 15$
0.1	3.338	3.338	3.338
0.3	4.302	4.319	4.323
0.5	5.424	5.792	6.018
0.7	6.062	7.195	8.393
0.9	6.042	7.818	10.276
1.1	5.622	7.663	10.850
1.3	5.071	7.099	10.417
1.5	4.534	6.425	9.565
1.7	4.061	5.784	8.656
1.9	3.660	5.223	7.830

In Table 8.3 we present the values of $E_\lambda\{T_S^{(1)}; k_1, k_2\}$.

8.2.3 Truncated SPRT of Type II

In type II truncation, the upper boundary is linear up to level n^*, and then stays constant at that level. The lower boundary is linear until it reaches the level n^*, i.e.

$$B_L^{(2)}(t) = I\{0 \le t \le n^* + k_1\}(-k_1 + t)^+,$$
$$B_U^{(2)}(t) = I\{0 \le t \le n^* - k_2\}(k_2 + t) + I\{n^* - k_2 \le t \le n^* + k_1\}n^*. \qquad (8.14)$$

The corresponding stopping times are

$$T_L^{(2)} = \inf\{t > 0 : N(t) = -k_1 + t\}, \quad T_U^{(2)} = \inf\{t > 0 : N(t) \ge \min(k_2 + t, n^*)\}$$
$$T_S^{(2)} = \min(T_L^{(2)}, T_U^{(2)}).$$

The region between the two parallel boundary lines is called the *continuation region*. Define the defective density

$$g_S^{(2)}(j, t; k_1, k_2) = P\{N(t) = j, T_S^{(2)} > t\}. \qquad (8.15)$$

Let $t^* = n^* + k_1$. This is the last t-coordinate in the continuation region. In order to compute $g_S^{(2)}(j, t; k_1, k_2)$ we partition the continuation region to blocks. The initial block is $B_0 = \{(t, j) : 0 \le t \le k_1; 0 \le j < k_2 + t\}$. Notice that $B_U^{(2)}(k_1) = k^* = k_1 + k_2$. Moreover, we assume that $n^* = l^*k^*$ where l^* is a positive integer. We partition the continuation region, from $t = k_1$ till $t = t^*$, in addition to the initial block B_0, to l^* additional block, B_l, $l = 1, \ldots l^*$, where $B_l = \{(t, j) : k_1 + (l-1)k^* \le t \le k_1 + lk^*; -k_1 + t < j < k_2 + t - 1\}$.

Let $t_l = k_1 + lk^*$ denote the right-hand limit of the t-interval of the l-th block.

8.2.3.1   Operating Characteristics

The upper boundary is $k_2 + t$. The lower boundary is 0. Thus, if $j \le k_2$, then $g_S^{(2)}(j, t; k_1, k_2) = p(j, \lambda t)$. The case of $k_2 = 0$ requires a special derivation (see Zacks 1991). Consider the region below the boundary $B(t) = t$. For a point $(j + l, j)$, $l = 1, 2, \ldots$ in this region

$$g_0(j, j + l) = I\{j = 0\}p(0, \lambda l)$$

$$+ I\{j > 0\} \left[p(j, \lambda(j + l)) - \sum_{i=0}^{j-1} p(i, \lambda i)g_0(j - i, j + l - i) \right].$$
$$(8.16)$$

This difference equation can be solved by the method of generating functions (see Zacks 2017).

The solution is

$$g_0(j, j + l) = \sum_{i=0}^{j} p(i, \lambda(i + l))p(j - i, \lambda(j + l - i)). \qquad (8.17)$$

Finally, for $k_2 < j < k_2 + l - 1$, and $l > 1$,

$$g_S^{(2)}(j, j + l; k_1, k_2) = p(j, \lambda(j + l))$$

$$- \sum_{i=1}^{j-k_2} p(k_2 + i, \lambda i)g_0(j - k_2 - i, j + l - i). \qquad (8.18)$$

(1) In Block B_1:

All these blocks are similar parallelograms between the two parallel boundaries. We therefore consider at the beginning the first parallelogram, block B_1, between the vertices A, B, C, and D, where $A = (k_1, 0)$, $B = (k_1, k^*)$, $C = (k_1 + k^*, k^*)$, and $D = (k_1 + k^*, 2k^*)$. The formulas for other blocks will be displayed later.

Thus, block B_1 can be partitioned to a lower sub-block, the right triangle ABC, and an upper sub-block, the right triangle BCD. The Poisson sample path $N(t)$ crosses from block B_0 to the lower sub-block of B_1 at some random level $m = 1, 2, \ldots, k^* - 1$, with probability $g_S^{(2)}(m, k_1; k_1, k_2)$. Since the sample path of $N(t)$ is nondecreasing, the sample path can cross the lower boundary of the block only at levels $m, m + 1, \ldots, k^* - 1$, with conditional probabilities

$$P\left\{T_L^{(2)} = k_1 + l,\, T_L^{(2)} < T_U^{(2)} \mid N(k_1) = m\right\} = I\{l \geq m\} f_L(l;\, m,\, k^* - m).$$
(8.19)

Notice that $f_L(l;\, m,\, k^* - m) = 0$ for $l < m$. It follows that

$$P\{T_L^{(2)} = k_1 + l,\, T_L^{(2)} < T_U^{(2)}\}$$

$$= \sum_{m=1}^{k^*-1} g_S^{(2)}(m, k_1;\, k_1, k_2) f_L(l;\, m,\, k^* - m)$$
(8.20)

$$P\{T_L^{(2)} = k_1\} = \exp\{-k_1\lambda\}.$$

We also define

$$\pi_1^{(2)}(\lambda) = \sum_{l=0}^{k^*-1} P\left\{T_L^{(2)} = k_1 + l,\, T_L^{(2)} < T_U^{(2)}\right\}.$$
(8.21)

(3) In block $B_l, l = 2, \ldots, l^*$:

As before, let $t_l = k_1 + l,\, l = 0, 1, \ldots$. If the sample path of $N(t)$ crosses the line BC before crossing the lower boundary, it might then cross the upper boundary, or cross to the next block.

It is easy to compute $g_S^{(2)}(j, t_l;\, k_1, k_2)$ recursively, starting with $g_S^{(2)}(j, k_1;\, k_1, k_2)$. Thus, for every $l = 0, 1, \ldots, (l^* - 2)k^*$

$$g_S^{(2)}(j, t_{l+1};\, k_1, k_2) = \sum_{i=l+1}^{\min\{j, k^*+l-2\}} g_S^{(2)}(j, t_l;\, k_1, k_2) p(j - i, \lambda).$$
(8.22)

Recall that $p(-j, \lambda) = 0$ for all $j > 0$. The l-th block starts at $(t_{(l-1)k^*}, (l-1)k^*)$ and ends at (t_{lk^*}, lk^*). If the sample path of $N(t)$ reaches the l-th block without stopping, then

$$P\left\{T_L^{(2)} = t_{(l-1)k^*} + j,\, T_L^{(2)} < T_U^{(2)} \mid N(t_{(l-1)k^*}) = (l - 1)k^* + m\right\}$$
$$= I\{j \geq m\} f_L(j;\, m,\, k^* - m)$$
(8.23)

and

$$P\left\{T_L^{(2)} = t_{(l-1)k^*} + j,\, T_L^{(2)} < T_U^{(2)}\right\}$$
$$= \sum_{m=1}^{k^*-1} g_S^{(2)}((l - 1)k^* + m, t_{(l-1)k^*};\, k_1, k_2) f_L(j;\, m,\, k^* - m).$$
(8.24)

Table 8.4 Values of $\pi(\lambda)$ for the case of $k_1 = 3$, $k_2 = 7$, and $n^* = 10$

λ	$\pi(\lambda)$
0.1	1.0000
0.3	0.9987
0.5	0.9587
0.7	0.7896
0.9	0.5241
1.1	0.2884
1.3	0.1397
1.5	0.0631
1.7	0.0279

Furthermore,

$$\pi_l(\lambda) = \sum_{j=0}^{k^*-1} P\left\{ T_L^{(2)} = t_{(l-1)k^*} + j, \, T_L^{(2)} < T_U^{(2)} \right\}. \tag{8.25}$$

Finally, the operating characteristic is

$$\pi(\lambda) = \sum_{l=1}^{l^*} \pi_l(\lambda). \tag{8.26}$$

In Table 8.4, we present the values of $\pi(\lambda)$ for the case of $k_1 = 3, k_2 = 7$, $n^* = 10$.

The expected value of $T_S^{(2)}$ is

$$E_\lambda\{T_S^{(2)}\} = \int_0^{l^* k^*} P\{T_S^{(2)} > t\}dt$$
$$= \int_0^{k_1} P_{k_2}(t, \lambda)dt + \sum_{l=1}^{(l^*-1)k^*-1} \int_{t_1}^{t_{l+1}} P_\lambda\{T_S^{(2)} > t\}dt, \tag{8.27}$$

where

$$P_{k_2}(t, \lambda)$$
$$= I\{t \le k_1\}\left[P(k_2, \lambda) + \sum_{j=k_2+1}^{k+\lfloor t \rfloor} \sum_{i=0}^{j} g_S^{(s)}(i, \lfloor t \rfloor; k, k)p(j - i, (t - \lfloor t \rfloor)\lambda) \right] \tag{8.28}$$

and

$$P_\lambda\{T_S^{(2)} > t\} = \sum_{l=k_1}^{(l^*-1)k^*-1} I\{t_l < t < t_{l+1}, k_1 < l < (l^* - 1)k^*\}$$

$$\left[\sum_{j=l-k_1+1}^{l+k_2-1} g_S^{(2)}(j, t_l; k_1, k_2) P(k_2 + l - 1 - j, \lambda(t - t_l)) \right]$$

$$+ I\{(l^* - 1)k^* < t < l^*k^*\} \sum_{\lfloor t \rfloor - k_1+1}^{m-1} g_S^{(2)}(j, \lfloor t+\rfloor; k_1, k_2)$$

$$P(m - 1 - j, \lambda(t - \lfloor t \rfloor)).$$

After some integrations we get

$$E_\lambda\{T_S^{(2)}\} = \frac{1}{\lambda} \left[\sum_{i=0}^{k_2} (1 - P(i, \lambda)) \right.$$

$$+ \sum_{l=2}^{k_1} \sum_{j=1}^{l+k_2-1} g_S^{(2)}(j, l; k_1, k_2) \sum_{i=0}^{l+k_2-1-j} (1 - P(i, \lambda))$$

$$+ \sum_{l=k_1+1}^{(l^*-1)k^*} \sum_{j=l-k_1+1}^{l+k_2-1} g_S^{(2)}(j, l; k_1, k_2) \sum_{i=0}^{l+k_2-1-j} (1 - P(i, \lambda))$$

$$+ \sum_{l=(l^*-1)k^*+1}^{l^*k^*-2} \sum_{j=l-k_1+1}^{m-1} g_S^{(2)}(j, l; k_1, k_2) \sum_{i=0}^{m-1-j} (1 - P(i, \lambda)) \left. \right].$$

$$\tag{8.29}$$

To conclude this section, we present a comparison between the characteristic functions of the two types of truncated sequential designs. For this purpose, consider a case where the two hypotheses are $H_0 : \beta \geq 2000$, against $H_1 : \beta \leq 1000$ h. For the error probabilities we take $\varepsilon_1 = \varepsilon_2 = 0.05$. The SPRT has two parallel line boundaries, with slope $b = 0.0007213$ and intercepts -4.25 and 4.25. Making the transformation to $\mu = \lambda/b$ we obtain $\mu_0 = 0.6932$ and $\mu_1 = 1.3863$. In Table 8.5 we present the values of $\pi(\lambda)$ and $E\{T\}$ for the two types of truncation.

We see in this table that the $\pi(\mu)$ values are very similar, but the expected length of the trials are significantly shorter in type II than in type I.

Table 8.5 Parameters of type I are $k_1 = 3, k_2 = 40$; parameters of type II are $k_1 = 3, k_2 = 7, m = 40$

β	μ	$\pi^{(1)}(\mu)$	$\pi^{(2)}(\mu)$	$E\{T_S^{(1)}\}$	$E\{T_S^{(2)}\}$
2000	0.7	0.987	0.958	9.764	9.657
1386.4	1.0	0.642	0.650	20.782	18.198
1000	1.4	0.116	0.112	25.134	14.358
866.5	1.6	0.046	0.044	23.372	10.745
603.2	2.0	0.008	0.008	19.345	6.614

8.3   Sequential Estimation

In sequential estimation procedures one starts with an initial sample of size k and then decides, according to an accuracy criterion, whether to stop or continue sampling. Sampling after the initial sample is done one by one, sequentially, until a stopping boundary is reached. In the previous section we discussed the sequential testing hypotheses about the intensity λ of a Poisson process. In the present section we will present the sequential estimation of the mean β of an exponential distribution. This is the continuous version of the Poisson process, where the exponential distribution is the distribution of the inter-arrival times in the Poisson process. We present here the paper of Zacks and Mukhopadhyay (2006).

Let $\{X_n, n \geq 1\}$ be a sequence of i.i.d. random variables having an exponential distribution with mean β. Given a sample of n observations, the sample mean $\bar{X}_n = \frac{1}{n}\sum_{i=1}^{n} X_i$ is a minimal sufficient statistic. A boundary sequence $B_n = f(\bar{X}n)$ is determined according to some criterion, and correspondingly a stopping variable $N = \min\{n \geq k : n \geq B_n\}$ is prescribed. At stopping, the parameter β is estimated by $\hat{\beta}_N = \bar{X}_N$. The corresponding loss function is $L(\hat{\beta}_N, \beta)$, and the stopping rule is designed to minimize the corresponding risk $E\{L(\hat{\beta}_N, \beta)\}$. Starr and Woodroofe (1972), Woodroofe (1977, 1982), and Ghosh et al. (1997) considered the loss function $L(\hat{\beta}_n, \beta) = A(\hat{\beta}_n - \beta)^2 + cn$, which led to the stopping variable

$$N^{(1)} = \min\left\{n \geq k : n \geq \left(\frac{A}{c}\right)^{1/2} \bar{X}_n\right\}. \tag{8.30}$$

Datta and Mukhopadhyay (1995) introduced the stopping rule

$$N^{(2)} = \min\left\{n \geq k : n \geq \left(\frac{A}{W}\right) \bar{X}_n^2\right\}, \tag{8.31}$$

to obtain an estimator with a bounded risk w. In the following we derive the exact distributions of these stopping rules, and we compare their minimal risks.

8.3.1   Distributions of Stopping Times

Let $\{N(t), t \geq 0\}$ be a Poisson process with intensity $\lambda = \frac{1}{\beta}$ Since the inter-arrival times of this Poisson process are distributed like i.i.d. exponential random variables with mean β, we can translate the above stopping variables to stopping times with continuous boundaries. Let $\{\tau_n, n \geq 1\}$ be the random epochs of jumps of the sample path $N(t)$, i.e., $N(\tau_j) = j$. Since $S_n = \sum_{i=1}^{n} X_i \sim \beta G(n, 1)$, the pair (n, S_n) corresponds to $(N(\tau_n), \tau_n)$. Accordingly, stopping variable (Sect. 8.3.1) can be translated to the stopping time

$$T^{(1)} = \inf\{t \geq t_k : N(t) \geq v_1 t^{1/2}\}, \tag{8.32}$$

where $v_1 = \left(\frac{A}{c}\right)^{1/4}$ and $t_k = \left(\frac{k}{v_1}\right)^2$. Notice that $N^{(1)} = N(T^{(1)})$ and $\bar{X}_{N^{(1)}} = \frac{T^{(1)}}{N(T^{(1)})}$. Similarly,

$$T^{(2)} = \inf\{t \geq t_k : t \geq v_2 t^{2/3}\}, \tag{8.33}$$

where $v_2 = \left(\frac{A}{c}\right)^{1/3}$ and $t_k = \left(\frac{k}{v_1}\right)^{3/2}$. We realize also that these two boundaries are of the form $B(t) = vt^\alpha$, where $0 < v < \infty, 0 < \alpha < 1$. Thus $B(t)$ is a concave boundary.

8.3.2 Distribution of First Crossing Concave Boundaries

We consider here a general stopping time of the form $T = \inf\{t \geq t_k : N(t) \geq \gamma t^\alpha\}$. It is easy to prove that $P_\lambda\{T < \infty\} = 1$. Indeed

$$P_\lambda\{T = \infty\} = \lim_{t \to \infty} P_\lambda\{T > t\}$$
$$= \lim_{t \to \infty} P_\lambda\left\{\bigcap_{s \leq t} N(s) < \gamma t^\alpha\right\} \leq \lim_{t \to \infty} P_\lambda\{N(t) < \gamma t^\alpha\} = 0, \tag{8.34}$$

since $\lim_{t \to \infty} \frac{N(t)}{t} = \lambda$ a.s. (see Gut 1988, p. 83).

As before, let $g(j, t) = P\{N(t) = j, T > t\}$. Also, let k be an integer greater than γ. For $l \geq k$, let $t_l = (l/\gamma)^{1/\alpha}$. Thus, the identity $\{N(t_k) = k\} = \{S_k = t_k\}$ yields according to the Poisson–Gamma relationship

$$P_\lambda\{T = t_k\} = P_\lambda\{N(t_k) \geq k\} = P_\lambda\{G(k, 1) \leq \lambda t_k\} = 1 - P(k-1, \lambda t_k). \tag{8.35}$$

Furthermore, since the stopping rule does not stop before time t_k,

$$g(j, t_k) = I\{0 \leq j \leq k - 1\} p(j, \lambda t_k). \tag{8.36}$$

For $l = k + 1, k + 2, \ldots$ we obtain the values of $g(j, t_l)$ recursively, in the following

$$g(j, t_l) = I\{j < l\} \sum_{i=0}^{\min\{j, l-1\}} g(i, t_{l-1}) p(j - i, \lambda). \tag{8.37}$$

Generally,

$$g(j, t) = \sum_{l=k}^{\infty} I\{t_l < t \leq t_{l+1}\} \sum_{i=0}^{\min\{j, l\}} g(i, t_l) p(j - i, \lambda(t - t_l)). \tag{8.38}$$

Finally,

$$P_\lambda\{T > t\} = \sum_{j < B(t)} g(j, t)$$

$$= \sum_{l=k+1}^{\infty} I\{t_{l-1} < t < t_l\} \sum_{j=0}^{l-2} g(j, t_{l-1}) P(l-1-j, \lambda(t-t_{l-1})).$$

(8.39)

Notice that the function $P\{T > t\}$ is continuous in t on (t_k, ∞). The p.d.f. of N is obtained recursively for all $n > k$, where

$$P\{N = k\} = 1 - P(k-1, \lambda t_k),$$

$$P\{N = n\} = P\{N > n-1\} - P\{N > n\}$$

$$= \sum_{j=0}^{n-2} g(j, t_{n-1}) - \sum_{j=0}^{n-1} g(j, t_n).$$

The density function of T is

$$f_T(t) = \lambda \sum_{l=k+1}^{\infty} I\{t_{l-1} < t < t < l\} \sum_{j=0}^{n-2} g(j, t_{n-1}) p(l-1-j, \lambda(t-t_{l-1})).$$

(8.40)

With this density one can compute the moments of the stopping time T, and the corresponding moments of N and of $\bar{X}_N$. The r-th moments of these are

$$E\{N^r\} = \sum_{n=k}^{\infty} n^r P\{N = n\}$$

$$= k^r (1 - P(k-1, \lambda t_k))$$

$$+ \sum_{n=k+1}^{\infty} n^r \left[\sum_{j=0}^{n-2} g(j, t_{n-1}) (1 - P(n-1, \lambda(t_n - t_{n-1}))) \right.$$

(8.41)

$$\left. + \exp\{-\lambda(t_n - t_{n-1})\} g(n-1, t_{n-1}) \right]$$

$$E\{\bar{X}_N^r\}$$

$$= \frac{r! \sum_{l=k+1}^{\infty} \left(\frac{t_{l-1}}{l}\right)^r \sum_{m=0}^{l-2} g(m, t_{l-1}) \sum_{j=0}^{r} \binom{l-m+j-1}{j}(1 - P(l-m+j-1, \lambda(t_1 - t_{l-1})))}{(\lambda t_{l-1})^j (r-j)!}.$$

(8.42)

Table 8.6 Exact values of $E\{N^{(1)}\}$ and $E\{\hat{\beta}_{i,N^{(1)}}\}$, for $A = 10$, $k = 3$, and $\beta = 1$

c	$E\{N^{(1)}\}$	$E\{\hat{\beta}_{1,N^{(1)}}\}$	$E\{\hat{\beta}_{2,N^{(1)}}\}$	$E\{\hat{\beta}_{3,N^{(1)}}\}$	$E\{\hat{\beta}_{4,N^{(1)}}\}$
0.5	4.712	0.8663	1.0992	1.0554	1.1106
0.1	9.482	0.8757	0.9800	0.9482	0.9737
0.05	13.472	0.9015	0.9744	0.9526	0.9706
0.01	31.882	0.9597	0.9916	0.9828	0.9910
0.005	44.305	0.9742	0.9967	0.9907	0.9964

8.3.3   Estimating β Under $N^{(1)}$

The estimator mentioned earlier is $\hat{\beta}_{1,N^{(1)}} = \bar{X}_{N^{(1)}}$. This estimator is biased. Mukhopadhyay and Cicconetti (2004) suggested three other estimators in order to reduce the bias. These are

$$\hat{\beta}_{2,N^{(1)}} = \left(\frac{N^{(1)}}{N^{(1)} - 1}\right) \bar{X}_{N^{(1)}}, \tag{8.43}$$

$$\hat{\beta}_{3,N^{(1)}} = \left(\frac{c}{A}\right)^{\frac{1}{2}} N^{(1)},$$

$$\hat{\beta}_{3,N^{(1)}} = \left(\frac{c}{A}\right)^{\frac{1}{2}} (N^{(1)} + 0.25497).$$

In Table 8.6 we present numerical values, for the expected values of $N^{(1)}$ and $\hat{\beta}_{i,N^{(1)}}$, $i = 1, \ldots, 4$.

8.4   Reliability Analysis for Damage Processes

Bogdanoff and Kozim (1985) define cumulative damage as the irreversible accumulation of damage throughout life, which ultimately leads to failure. In a series of papers Zacks (2004, 2010) modeled such failure time distributions, which result from cumulative damage processes. The amount of damage to a system at time t is modeled as a random process $\{Y(t), t > 0\}$, where $Y(t) \geq 0$ is a nondecreasing process, with $Y(t) \to \infty$ a.s. as $t \to \infty$. The failure time of a system subjected to such a damage process is the first instant at which $Y(t) \geq \beta$, where $0 < \beta < \infty$ is a threshold specific to the system. Thus, the failure distribution is a stopping time distribution. We present in this section the methodology of deriving such distribution and estimating its parameters. In particular we focus our attention here on a nonhomogeneous Poisson damage process, having an intensity function of the Weibull type, i.e., $\lambda(t) = (\lambda t)^{\nu}$, where $0 < \lambda, \nu < \infty$, compounded with a sequence of positive random damage amounts $\{X_n, n \geq 1\}$.

8.4.1   Compound Cumulative Damage Processes

The system is subjected to shocks at random times $0 < \tau_1 < \tau_2 < \ldots$, following a nonhomogeneous Poisson process, with intensity $\lambda(t)$. The amount of damage to the system at the n-th shock is a random variable $X_n, n \geq 1$. We assume that $X_1, X_2, \ldots$ are i.i.d. and that $\{X_n, n \geq 1\}$ are independent of $\{\tau_n, n \geq 1\}$. Let $N(0) = 0$ and $N(t) = \max\{n : \tau_n \leq t\}$. The process $\{N(t), t \geq 0\}$ has independent increments such that, for any $0 \leq s < t < \infty$,

$$P\{N(t) - N(s) = n\} = \exp\{-(m(t) - m(s))\}(m(t) - m(s))^n/n! \qquad (8.44)$$

where $n = 0, 1, \ldots$, and $m(t) = \int_0^t \lambda(s)ds$. The compound damage process is defined as

$$Y(t) = \sum_{n=0}^{N(t)} X_n. \qquad (8.45)$$

We restrict attention to compound Weibull process, in which $\lambda(t) = \lambda\nu(\lambda t)^{\nu-1}$. We assume also that the distribution F of X is absolutely continuous, with density f. The c.d.f. of $Y(t)$ is

$$D(y, t) = \sum_{n=0}^{\infty} \exp(-m(t)) \left(\frac{m(t)^n}{n!}\right) F^{(n)}(y), \qquad (8.46)$$

and the corresponding density is

$$d(y, t) = \sum_{n=1}^{\infty} \exp(-m(t)) \left(\frac{m(t)^n}{n!}\right) f^{(n)}(y). \qquad (8.47)$$

Here $m(t) = (\lambda t)^\nu$, and $\exp\{-m(t)\}(m(t)^n/n!) = p(n, (\lambda t)^\nu)$.

In the special case where X is distributed like an Exponential with mean $\frac{1}{\mu}$ we have $f^{(n)}(y) = \mu p(n-1, \mu y)$ and $F^{(n)}(y) = 1 - P(n-1, \mu y)$. A cumulative damage failure time is the stopping time

$$T(\beta) = \inf\{t > 0 : Y(t) \geq \beta\}. \qquad (8.48)$$

Accordingly,

$$P\{T(\beta) > t\} = D(\beta, t), 0 < t < \infty. \qquad (8.49)$$

This is the reliability function of the system. Hence, in the case of Weibull/Exponential damage process, to be designated as CPW/E

$$P\left(T(\beta) > t\right) = 1 - \sum_{n=0}^{\infty} p\left(n, (\lambda t)^{\nu}\right) P(n-1, \mu\beta)$$

$$= \sum_{j=0}^{\infty} p(j, \varsigma) P(j, ((\lambda t)^{\nu}), \tag{8.50}$$

where $\varsigma = \mu\beta = \beta/E\{X_1\}$. As mentioned before, the function (8.50) is the reliability of the system, as a function of t. We denote this reliability by $R(t; \lambda, \nu, \varsigma)$. This is a continuous, decreasing function of t, from $R(0; \lambda, \nu, \varsigma) = 1$ to $\lim_{t\to\infty} R(t; \lambda, \nu, \varsigma) = 0$. Notice that the density of $T(\beta)$ is

$$f_T(t; \beta) = -\frac{d}{dt} P\{T(b) > t\} = \lambda\nu(\lambda t)^{\nu-1} \sum_{j=0}^{\infty} p(j, \varsigma) p(j, (\lambda t)^{\nu}). \tag{8.51}$$

Hence, the hazard function is

$$h(t; \lambda, \nu, \varsigma) = \frac{\lambda\nu(\lambda t)^{\nu-1} \sum_{j=0}^{\infty} p(j, \varsigma) p(j, (\lambda t)^{\nu})}{\sum_{j=0}^{\infty} p(j, \varsigma) P(j, \lambda t)^{\nu})}. \tag{8.52}$$

The m-th moment of $T(\beta)$ is

$$E\{T^m(\beta)\} = \left(\frac{1}{\lambda^m}\right) \sum_{j=0}^{\infty} p(j, \varsigma) \frac{\Gamma\left(j + 1 + \frac{m}{\nu}\right)}{\Gamma(j+1)}. \tag{8.53}$$

Thus, moments of all orders exist.

8.4.2 Estimation of Parameters

Let $T_1, T_2, \ldots, T_n$ be i.i.d. random failure times, following a CPW/E process. The likelihood function of the parameters $(\lambda, \nu, \varsigma)$ is

$$L(\lambda, \nu, \varsigma; T_1, \ldots, T_n) = \lambda^{n\nu} \nu^n \left(\prod_{i=1}^{n} T_i^{\nu-1}\right) \prod_{i=1}^{n} \sum_{j=0}^{\infty} p(j, \varsigma) p(j, (\lambda T_i)^{\nu}). \tag{8.54}$$

Accordingly, the minimal sufficient statistic is the order statistic $(T_{(1)} < \cdots < T_{(n)})$. It is difficult to estimate all the three parameters simultaneously. We will simplify first by assuming that the parameter ν is known. Assuming that $\nu = 1$, it is straightforward to estimate (λ, ς) by the method of moments equations, since the moments formula (8.52) is relatively simple.

8.4.2.1   The Method of Moments Equations

The method of moments equations yields the two equations

$$\hat{\lambda} M_1 = 1 + \hat{\varsigma} \tag{8.55}$$

and

$$\hat{\lambda}^2 M_2 = 2 + 4\hat{\varsigma} + \hat{\varsigma}^2, \tag{8.56}$$

where $M_1 = \frac{1}{n} \sum_{i=1}^{n} T_i$ and $M_2 = \frac{1}{n} \sum_{i=1}^{n} T_i^2$ are the first two empirical moments.
From Eq. (8.55) we obtain the formula

$$\hat{\lambda} = \frac{1 + \hat{\varsigma}}{M_1}. \tag{8.57}$$

Substituting (8.57) in (8.56), we obtain a quadratic formula for ς, which is

$$\left(\frac{M_2}{M_1^2 - 1} \right) \hat{\varsigma}^2 + 2\hat{\varsigma} \left(\frac{M_2}{M_1^2 - 2} \right) + \left(\frac{M_2}{M_1^2 - 2} \right) = 0. \tag{8.58}$$

Notice that $\frac{M_2}{M_1^2 - 1} = \frac{S_T^2}{M_1^2} = C_T^2$, where C_T is the coefficient of variation. Thus, the quadratic equation is

$$C_T^2 \hat{\varsigma}^2 + 2\hat{\varsigma}(C_T^2 - 1) + C_T^2 - 1 = 0.$$

A positive real solution exists if $C_T < 1$. It is then

$$\hat{\varsigma} = \frac{(1 - C_T^2) + \sqrt{(1 - C_T^2)}}{C_T^2}. \tag{8.59}$$

In Table 8.7 we present simulation estimates of the $E\{\hat{\varsigma}\}$ and $E\{\lambda\}$, as well as their STD. The simulation is based on 100 independent runs.

Table 8.7 Simulation estimates of the expected MME and their STD, $\beta = 10$

λ	μ	n	$E\{\hat{\varsigma}_n\}$	$STD\{\hat{\varsigma}_n\}$	$E\{\hat{\lambda}_n\}$	$STD(\hat{\lambda}_n)$
1	1	25	10.7282	3.8327	1.1249	0.4439
		50	10.4728	2.5304	1.0684	0.2392
		100	10.2254	1.8645	1.0129	0.1588
1	3	25	34.1735	10.9083	1.1572	0.4291
		50	30.9005	7.3244	1.0491	0.2325
		100	30.6277	4.0550	0.9955	0.1505

8.4.2.2   Maximum Likelihood Estimator of ς When (λ, ν) Are Known

We develop now the maximum likelihood estimator (MLE) of ς, for $\lambda = 1$, $\nu = 1$. The log-likelihood function, given the sample of $T_1 \ldots T_n$, is

$$l(\varsigma; \mathbf{T}^{(n)}) = \sum_{i=1}^{n} \log E_\varsigma\{p(J, \lambda T_i)\}. \tag{8.60}$$

Notice that, $\frac{\partial}{\partial \varsigma} p(J, \varsigma) = p(J - 1, \varsigma) - p(J, \varsigma)$, where $J \sim Pois(\varsigma)$. Accordingly, the score function is, by the dominated convergence theorem,

$$\frac{\partial}{\partial \varsigma} l(\lambda, \varsigma : \mathbf{T}^{(n)}) = -n + \sum_{i=1}^{n} \frac{E_\varsigma\{p(J + 1, \lambda T_i)\}}{E_\varsigma\{p(j, \lambda T_i)\}} = 1. \tag{8.61}$$

The MLE is obtained by equating the score function to zero and solving for $\hat{\varsigma}$, the equation

$$\frac{1}{n} \sum_{i=1}^{n} \frac{E_\varsigma\{p(J + 1, \lambda T_i)\}}{E_\varsigma\{p(j, \lambda T_i)\}} = 1. \tag{8.62}$$

The solution of this equation was performed by numerical search. For example, for $\lambda = 1$, $\mu = 2$, $\beta = 10$, the true value of ς is 20. A random sample of $n = 100$ values of T yields the MLE $\hat{\varsigma} = 20.1$. Ten such replicas yield the following estimates $E\{\hat{\varsigma}\} = 18.65$, and $STD\{\hat{\varsigma}\} = 1.3344$. As expected, the MLE of ς has a smaller standard deviation than that for low level, and MME. The numerical search takes long computation time.

Chapter 9
Random Choice of Fractional Replications

In the present chapter several papers on random choice of fractional replications, which were published by Ehrenfeld and Zacks (1961, 1963, 1967) and Zacks (1963a,b, 1964, 1968) will be discussed. The first two sections introduce classical notions and notations in order to facilitate the material in the following sections. The random choice procedures presented here were designed to eliminate the bias (aliases) in estimating parameters in the classical fractional replication designs. This bias is, as will be shown, a linear function of the aliases parameters. The variances of the estimated parameters increase as a result of the random choice of fractional replications. The question is whether the mean-squared-errors of the estimators are reduced. We discuss also optimal strategies, generalized LSE, and randomized fractional weighing designs. The approach in the present chapter is a marginal analysis compared to a conditional analysis in the classical treatment of fractional replications. The difference between the two approaches is similar to the difference between the design and the modeling approaches in sampling surveys. The design approach is to choose the units of a population at random, and the properties of estimators depend on the randomization procedure. In the modeling approach the analysis is Bayesian, conditional on the units chosen, not necessarily at random. The population units are the fractions (blocks) of the full factorial experiment.

9.1 Models for Factorial Experiments

Factorial experiments are designed to assess the effects of controllable factors on the yield of a product, the interactions between different factors, and more. The reader is referred to Chapter 11 of Kenett and Zacks (2014, second edition) for a textbook explanation of the material.

Suppose that in a given experiment m factors, labeled $A_1 \ldots, A_m$ are specified. Let a_j denote the number of levels at which factor A_j is applied. If it is expected

© Springer Nature Switzerland AG 2020
S. Zacks, *The Career of a Research Statistician*, Statistics for Industry,
Technology, and Engineering, https://doi.org/10.1007/978-3-030-39434-9_9

that the effects of a factor, in the feasible experimental range, is linear, then two levels are sufficient. For nonlinear effects three or more levels are required. A p^m factorial experiment is one in which all factors are tested at p levels. We restrict the discussion in the present chapter to the cases of $p = 2$ or $p = 3$. The levels for cases of $p = 2$ are denoted by 0 for low level and 1 for high level. In the case of $p = 3$ the levels are 0, 1, 2. We restrict attention now to the case of $p = 2$ (linear effects). The case of $p = 3$ (quadratic effects) will be discussed later.

9.1.1   The Linear Case, $p = 2$

A treatment combination is designated by a vector $(i_1, \ldots, i_m)$, where $i_j = 0, 1$ for all $j = 1, 2, \ldots, m$. The treatment combinations are ordered from 0 to $2^m - 1$, according to the binary expansion $v = \sum_{j=1}^m i_j 2^{j-1}$. The yield of a treatment combination is Y_v given by the linear model

$$Y_v = \sum_{l=0}^{2^m-1} c_{vl}^{(2^m)} \beta_l + \varepsilon_v, \, v = 0, \ldots, 2^m - 1. \tag{9.1}$$

The index l of the component β is $l = \sum_{j=1}^m i_j 2^{j-1}$. The coefficients $c_{vl}^{(2^m)}$ are elements of the Hadamard matrix $C^{(2^m)}$ which is the m-fold Kronecker multiplication of $C^{(2)}$, iteratively

$$C^{(2)} = \begin{pmatrix} 1 & -1 \\ 1 & 1 \end{pmatrix} \tag{9.2}$$

and for $m = 2, 3, \ldots$

$$C^{(2^m)} = C^{(2)} \otimes C^{(2^{m-1})} = \begin{pmatrix} C^{(2^{m-1})} & -C^{(2^{m-1})} \\ C^{(2^{m-1})} & C^{(2^{m-1})}. \end{pmatrix} \tag{9.3}$$

Notice that the matrix $C^{(2^m)}$ is an orthogonal array of elements which are ± 1, with $C^{(2^m)\prime} C^{(2^m)} = 2^m I$, and $(C^{(2^m)})^{-1} = \frac{1}{2^m}(C^{(2^m)})'$. Let ε be a random vector satisfying $E\{\varepsilon\} = 0$, and $V[\varepsilon] = \sigma^2 I$. Thus, in matrix notation

$$\mathbf{Y} = (C^{(2^m)})\beta + \varepsilon. \tag{9.4}$$

The LSE of the regression coefficient β is

$$\hat{\beta} = \frac{1}{2^m}(C^{(2^m)})'\mathbf{Y}. \tag{9.5}$$

This is an unbiased estimator with variance–covariance matrix

$$V[\hat{\beta}] = \frac{\sigma^2}{2^m} I. \tag{9.6}$$

9.2 Fractional Replications

Full factorial experiments are feasible when the number of factors, m, is small. For example, if $m = 6$, the number of treatment combinations, each one at two levels, is $2^6 = 64$. Even such an experiment might be considered sometimes too big. However, there are many situations in industry, in which the number of factors is over $m = 20$. In such cases the number of treatment combinations in a full factorial is greater than $2^{20} = 1,048,576$. It might be impractical to make all these trials. Moreover, in such models the number of all the main effects and interactions (the dimension of β) is also that large, and one may be interested only in a smaller number of parameters. The method described below allows us to partition the 2^{20} treatment combinations into 2^{14} blocks of size 2^6. Each such block is a fractional replication. The experimenter can chose one or several of these fractional replications for estimating the parameters of interest. There are different methods for choosing the fractional replications. The method described here is the basis for the theory presented in following sections.

9.2.1 Classical Notation for Small Size 2^s Experiments

In Table 9.1 we show the relationship between the index $l = \sum_{j=1}^{m} i_j 2^{j-1}$ of the components of β and the main effects and interactions in a 2^5 factorial experiment. We use here for clarity classical notations for the effects. The mean effect is denoted by μ; the main effects by A, B, C, D, E; the interactions by AB, AC, etc.

From the matrix of coefficients $C^{(2^5)}$ we copy the column vectors corresponding to the five main effects. We denote the treatment combinations by small letters $(1)a, b, ab, \ldots$ where these stand for $a^{i_1} b^{i_2} c^{i_3} d^{i_4} e^{i_5}$, $i_j = 0, 1$. A zero power indicates low level of the factor, and power one indicates high level of the factor. Notice that the treatment combination (1) stands for all factors at low level. We keep the word short by indicating only the factors with high level. The following table presents the coefficients for the main effects and the generators.

The generators are high order interactions, according to which the partition to fractional blocks is done. In the example below (Table 9.2), the generators are the interaction parameters ABC, CDE. These two generators span a defining subgroup $\{\mu, ABC, CDE, ABDE\}$. Four blocks of eight treatment combinations (t.c.) are formed according to the entries in Table 9.2, under the columns of the generators. The blocks are denoted as $B_{--}, B_{-+}, B_{+-}, B_{++}$.

Table 9.1 Main effects and interactions

l	Binary	Effect	l	Binary	Effect
0	0, 0, 0, 0, 0	μ	16	0, 0, 0, 0, 1	E
1	1, 0, 0, 0, 0	A	17	1, 0, 0, 0, 1	AE
2	0, 1, 0, 0, 0	B	18	0, 1, 0, 0, 1	BE
3	1, 1, 0, 0, 0	AB	19	1, 1, 0, 0, 1	ABE
4	0, 0, 1, 0, 0	C	20	0, 0, 1, 0, 1	CE
5	1, 0, 1, 0, 0	AC	21	1, 0, 1, 0, 1	ACE
6	0, 1, 1, 0, 0	BC	22	0, 1, 1, 0, 1	BCE
7	1, 1, 1, 0, 0	ABC	23	1, 1, 1, 0, 1	$ABCE$
8	0, 0, 0, 1, 0	D	24	0, 0, 0, 1, 1	DE
9	1, 0, 0, 1, 0	AD	25	1, 0, 1, 1, 1	ADE
10	0, 1, 0, 1, 0	BD	26	0, 1, 0, 1, 1	BDE
11	1, 1, 0, 1, 0	ABD	27	1, 1, 0, 1, 1	$ABDE$
12	0, 0, 1, 1, 0	CD	28	0, 0, 1, 1, 1	CDE
13	1, 0, 1, 1, 0	ACD	29	1, 0, 1, 1, 1	$ACDE$
14	0, 1, 1, 1, 0	BCD	30	0, 1, 1, 1, 1	$BCDE$
15	1, 1, 1, 1, 0	$ABCD$	31	1, 1, 1, 1, 1	$ABCDE$

The four blocks of treatment combinations are:

$$B_{--} = \{(1), ab, acd, bcd, ace, bce, de, abde\}$$

$$B_{-+} = \{ac, bc, d, abd, e, abe, acde, bcde\}$$

$$B_{+-} = \{a, b, cd, abcd, ce, abce, ade, bde\}$$

$$B_{++} = \{c, abc, ad, bd, ae, be, cde, abcde\}.$$

We show now how to compute the expected value of the LSE of the parameters, given the vector of observations Y, of the block (fractional replication) observed. For illustration, we will take the block B_{--}. The coefficients of $C^{(2^5)}$ are read from Table 9.2. According to Table 9.2, the LSE of A is

$$\hat{A} = \frac{1}{8}[-Y(1) + Y(ab) + Y(acd) - Y(bcd) \tag{9.7}$$
$$+ Y(ace) - Y(bce) - Y(de) + Y(abde)].$$

The expected value of this estimator can be evaluated again with Table 9.2, assuming that the parameters of interest are: A, B, C, AB, AC, BC. Then appropriate coefficients are shown in Table 9.3.

Accordingly

$$E\{\hat{A}\} = A - BC. \tag{9.8}$$

We see that if $BC \neq 0$, then $\hat{A}$ is a biased estimator.

Table 9.2 Coefficients of main effects and of defining parameters

t.c.	μ	A	B	C	D	E	ABC	CDE
(1)	1	−1	−1	−1	−1	−1	−1	−1
a	1	1	−1	−1	−1	−1	1	−1
b	1	−1	1	−1	−1	−1	1	−1
ab	1	1	1	−1	−1	−1	−1	−1
c	1	−1	−1	1	−1	−1	1	1
ac	1	1	−1	1	−1	−1	−1	1
bc	1	−1	1	1	−1	−1	−1	1
abc	1	1	1	1	−1	−1	1	1
d	1	−1	−1	−1	1	−1	−1	1
ad	1	1	−1	−1	1	−1	1	1
bd	1	−1	1	−1	1	−1	1	1
abd	1	1	1	−1	1	−1	−1	1
cd	1	−1	−1	1	1	−1	1	−1
acd	1	1	−1	1	1	−1	−1	−1
bcd	1	−1	1	1	1	−1	−1	−1
abcd	1	1	1	1	1	−1	1	−1
e	1	−1	−1	−1	−1	1	−1	1
ae	1	1	−1	−1	−1	1	1	1
be	1	−1	1	−1	−1	1	1	1
abe	1	1	1	−1	−1	1	−1	1
ce	1	−1	−1	1	−1	1	1	−1
ace	1	1	−1	1	−1	1	−1	−1
bce	1	−1	1	1	−1	1	−1	−1
abce	1	1	1	1	−1	1	1	−1
de	1	−1	−1	−1	1	1	−1	−1
ade	1	1	−1	−1	1	1	1	−1
bde	1	−1	1	−1	1	1	1	−1
abde	1	1	1	−1	1	1	−1	−1
cde	1	−1	−1	1	1	1	1	1
acde	1	1	−1	1	1	1	−1	1
bcde	1	−1	1	1	1	1	−1	1
abcde	1	1	1	1	1	1	1	1

The biasing parameter(s) are called aliases. If all the four blocks are observed, all the 32 parameters can be estimated unbiasedly, a full factorial situation. Thus, if a block is chosen at random, with probability $\frac{1}{4}$, then all the parameters of interest can be estimated unbiasedly. Notice that these are two types of unbiasedness. Conditional unbiasedness due to the linear model is model unbiasedness. Unbiasedness due to random choice of a block (fractional replication) is design unbiasedness. See the different definitions in Chap. 6.

Table 9.3 Coefficients for estimating $E\{\hat{A}\}$

$\hat{A}$	A	B	C	AB	AC	BC
$-Y(1)$	1	1	1	-1	-1	-1
$Y(ab)$	1	1	-1	1	-1	-1
$Y(acd)$	1	-1	1	-1	1	-1
$-Y(bcd)$	1	-1	-1	1	1	-1
$Y(ace)$	1	-1	1	-1	1	-1
$-Y(bce)$	1	-1	-1	1	1	-1
$-Y(de)$	1	1	1	-1	-1	-1
$Y(abde)$	1	1	-1	1	-1	-1
$\sum$	8	0	0	0	0	-8

9.2.1.1   The Algebra of 2^s Fractional Replications

As in (9.3), for any $1 < s < m$,

$$C^{(2^m)} = C^{(2^{m-s})} \otimes C^{(2^s)}. \tag{9.9}$$

We present here some important results concerning the elements of these Hadamard matrices.

I. The elements of $C^{(2^m)}$ are related to those of $C^{(2^s)}$ and those of $C^{(2^{m-s})}$ according to

$$c^{(2^m)}_{(i+j)(2^s),l} = c^{(2^{m-s})}_{j,q_l} c^{(2^s)}_{i,r_l}, \quad \text{for all } j = 0, \ldots, 2^{m-s} - 1;$$

$$i = 0, \ldots, 2^s - 1; \; q_l = \lfloor \frac{l}{2^s} \rfloor; \; r_l = l - q_l 2^s. \tag{9.10}$$

II. The following result holds for $p = 2$, but does not necessarily holds for $p > 2$. In a 2^m factorial system, if v is the t.c. $(i_1, \ldots, i_m)$, $v = \sum_{j=1}^{m} i_j 2^{j-1}$, and $u = \sum_{j=1}^{m} l_j 2^{j-1}$ is the parameter order, then

$$c^{2^m}_{vu} = (-1)^k, \quad \text{where } k = \sum_{j=0}^{m-1} l_j (l - i_j). \tag{9.11}$$

The reader can demonstrate formula (9.11) in the case of $m = 5$ by using Table 9.2. For example, $c^{(2^5)}_{17,15} = (-1)^3$.

A set of parameters are called mutually independent if no parameter in the set can be obtained as a product of other parameters in the set. Accordingly, all main effects in a 2^m factorial are mutually independent. However, B depends on $\{A, AB\}$.

9.3 Randomly Chosen Fractional Replications

We discuss in the present section two types of randomly chosen fractional replications for observation and estimation. In the first type, called RP.I, a subgroup of 2^s parameters of interest is chosen. The 2^m treatment combinations are partitioned into 2^{m-s} blocks, each one containing 2^s treatment combinations. One or several blocks are chosen at random, for observation and estimation, from the list of the 2^{m-s} blocks. Another random choice procedure, called RP.II, partitions the treatment combinations to 2^s blocks, each containing 2^{m-s} treatment combinations, according to the subgroup of interesting parameters. n treatment combinations are chosen at random from each block, for observation and estimation. In the following we present the properties of these randomization procedures. The first type is a cluster sampling of blocks. The second type is stratified sampling within the blocks.

9.3.1 Randomization Procedure R.P.I.

The 2^{m-s} blocks are constructed in the following manner:

(i) A subgroup of 2^s interesting parameters is chosen and specified. Let $\boldsymbol{\alpha}$ denote the vector of corresponding parameters.

(ii) A set of $m - s$ parameters is chosen. These parameters serve as generators and should be disjoint from the interesting parameters and independent. Designate these parameters as $\{\beta_{d_0}, \beta_{d_1}, \ldots, \beta_{d_{m-s-1}}\}$.

(iii) Specify the subgroup of defining parameters, spanned by the generators in (ii). This subgroup is called the defining parameters.

(iv) Classify all the treatment combinations into 2^{m-s} blocks, following this rule:

$$\text{if t.c. } x \leftrightarrow (i_0, \ldots, i_{m-1}) \text{ for all } j = 0, 1, \ldots, m - s - 1,$$

$$\text{then } x \in X\nu \text{ where } \nu = \sum_{j=0}^{m-s-1} a_j 2^j. \tag{9.12}$$

This classification rule is equivalent to the one illustrated in Table 9.2. To simplify the notation, let $S = 2^s$, and $M = 2^{m-s}$. We will also assume that the vector of interesting parameters is $\boldsymbol{\alpha}' = (\beta_0, \ldots, \beta_{S-1})$. The generators of the defining parameters are the M main effects $(\beta_S, \beta_{2S}, \ldots, \beta_{(M-1)S})$. Let $\boldsymbol{\beta}'_{(u)} = (\beta_{Su}, \beta_{1+Su}, \ldots, \beta_{(u+1)S-1}), u = 1, \ldots, M - 1$, be the vector of parameters alias to the components of $\boldsymbol{\alpha}$. The statistical linear model is

$$Y(X_\nu) = \left(C^{(s)}\right)\boldsymbol{\alpha} + \sum_{u=1}^{M-1} c_{\nu u}^{(M)} \left(C^{(s)}\right) \boldsymbol{\beta}_{(u)} + \epsilon_\nu, \text{ for all } \nu = 0, \ldots, M - 1. \tag{9.13}$$

Let

$$(H_v) = (c_{v1}^{(M)}, c_{v2}^{(M)}, ..., c_{v(M-1)}^{(M)}) \otimes (C^{(S)}).$$ (9.14)

Notice also that due to the fact that the columns of $(C^{(M)})$ sum up to $(M, 0, 0, \ldots, 0)$.

$$\sum_{v=1}^{M-1} (H_v)\boldsymbol{\beta} = 0.$$ (9.15)

In the randomization procedure we choose at random, with or without replacement, $n \geq 1$ blocks of X_v. Let $\{V_i, i = 1, \ldots, n\}$ denote the indexes of the randomly chosen blocks. The corresponding LSE of $\boldsymbol{\alpha}$ is

$$\hat{\boldsymbol{\alpha}}_{\bar{V}} = \left(\frac{1}{nS}\right)\left(C^{(s)}\right)' \sum_{i=1}^{n} Y(X_{V_i})$$

$$= \left(\frac{1}{n}\right) \sum_{i=1}^{n} \left(\boldsymbol{\alpha} + \left(\frac{1}{S}\right)\left(C^{(S)}\right)'(H_{V_i})\boldsymbol{\beta}\right).$$ (9.16)

It follows according to (9.15) that the design mean is

$$E\{\hat{\boldsymbol{\alpha}}_{\bar{V}}\} = \boldsymbol{\alpha}.$$ (9.17)

The variance–covariance of this estimator, when $n = 1$, is

$$V[\hat{\boldsymbol{\alpha}}] = \left(\frac{\sigma^2}{S}\right) I^{(S)} + \sum_{u=1}^{M-1} \boldsymbol{\beta}_{(u)}\boldsymbol{\beta}_{(u)}'.$$ (9.18)

9.3.1.1   Randomization Procedure R.P.II

In randomization procedure R.P.II the treatment combinations are partitioned into S blocks of size M, where the defining parameters are the interesting ones. Random samples of n treatment combinations are drawn from each block, independently. This is similar to stratified sampling procedure. Without loss of generality, assume that the s generators are the first s main effects $\{\beta_{2j}, j = 0, \ldots, S - 1\}$. The corresponding S blocks of treatment combinations are the sets

$$X_i = \left\{x : c_i^{(2^m)}(x_{i+j2^s}) = c_i^{(2^s)}(x_i)\right\}, i = 0, 1, \ldots, S - 1; j = 0, \ldots, M - 1.$$ (9.19)

We remark that the classification according to (9.19) is equivalent to the regular procedure of confounding the S chosen parameters. Let S_i denote the set of n treatment combinations chosen at random from X_i, i.e.

$$S_i(x) = \{x_{i+j_{l_1},2^s}, \ldots, x_{i+j_{l_n},2^s}\}, i = 0, \ldots, S-1. \tag{9.20}$$

Let $\bar{Y}_i$ denote the mean of the corresponding n observed yields of the sampled t.c. in $S_i(x)$. Let $\mathbf{Y}'_{II} = (\bar{Y}_0, \ldots, \bar{Y}_{S-1})$ the vector of all these S_i sample means. The LSE of the vector $\boldsymbol{\alpha}_{II}$ of the parameters of interest is

$$\hat{\boldsymbol{\alpha}}_{II} = \left(\frac{1}{S}\right)\left(C^{(S)}\right)' \mathbf{Y}_{II}. \tag{9.21}$$

Let $\hat{\beta}_{l,II}$ denote the l-th component of (9.21). The conditional expectation of this estimator is

$$E\left\{\hat{\beta}_{l,II}\right\} = \beta_l + \left(\frac{1}{S}\right)\sum_{i=0}^{S-1}\sum_{r=0}^{S-1} c_{il}^{(S)} c_{ir}^{(S)} \left[\sum_{q=1}^{M-1} c_{i.q}^{(M)} \beta_{r+qS}\right], \tag{9.22}$$

where

$$c_{i.q}^{(M)} = n^{-1}\sum_{k=1}^{n} c_{j_{i_k}q}^{(M)}. \tag{9.23}$$

Finally, under the random choice of the treatment combinations, the estimator $\hat{\boldsymbol{\alpha}}_{II}$ is unbiased, with components variance

$$V\{\hat{\beta}_{l,II}\} = \frac{\sigma}{nS} + \left(\frac{1}{nSM}\right)\sum_{q=1}^{M-1}\sum_{i=0}^{S-1}\left(c_{il}^{(S)}\right)^2\left(c_{ir}^{(S)}\beta_{r+qS}\right)^2. \tag{9.24}$$

If we designate by $V\{\hat{\beta}_{l,I}\}$ the variance of $\hat{\beta}_l$ according to R.P.I. then we have

$$V\{\hat{\beta}_{l,II}\} = S^{-1}\sum_{l=0}^{S-1} V\{\hat{\beta}_{l,I}\}. \tag{9.25}$$

9.4 Testing Hypotheses in R.P.I

In the papers of Ehrenfeld and Zacks (1961, 1967) testing hypotheses about the significance of the chosen (interesting) parameters were set up in analysis of variance (ANOVA) schemes, in which the test statistics are F-like ratios. The problem, as elaborated in those papers, is that of determining the critical levels for given level(s) of significance. We start with presenting the theoretical large samples approach, and then show a possible boot-strapping method that could be applied.

As before, we denote the vector of chosen parameters by $\boldsymbol{\alpha}$. The number of blocks (fractional replicates) is $M = 2^{m-s}$. The i-th component of $\boldsymbol{\alpha}$ is α_i, $i = 0, \ldots, S-1$. Its LSE based on the v-th block is

$$\hat{\alpha}_{vi} = \alpha_i + \eta_{vi} + \varepsilon_{vi}^*,$$ (9.26)

where

$$\eta_{vi} = \sum_{u=1}^{M-1} C_{vu}^{(M)} \beta_{i+u2^s}, \, v = 0, \ldots, M - 1.$$ (9.27)

We assume that the errors are normally distributed, $\varepsilon_{vi}^* \sim N(0, \sigma^2/S)$. We wish to test the hypothesis

$$H_0^{(i)} : \alpha_i = 0, \, \boldsymbol{\beta}^* \text{ arbitrary } (i = 0, \ldots, S - 1),$$ (9.28)

where $\boldsymbol{\beta}^*$ is the vector of nuisance parameters.

Let $\mathbf{c}_v'$ be the v-th randomly chosen row vector of $(C^{(M)})$. Define an $Sx(M - 1)$ matrix $(B) = (\boldsymbol{\beta}_{(1)}, \boldsymbol{\beta}_{(2)}, \ldots, \boldsymbol{\beta}_{(M-1)})$, where $\boldsymbol{\beta}_{(j)}' = (\beta_{(j-1)S}, \beta_{(j-1)S+1}, \ldots, \beta_{jS})$. The conditional distribution of $\hat{\boldsymbol{\alpha}}_v$ is that of $N(\boldsymbol{\alpha} + \mathbf{n}_v, (\sigma^2/S)I^{(S)})$, in which

$$\mathbf{n}_v = (B)\mathbf{c}_v, \, v = 0, 1, \ldots, M - 1.$$ (9.29)

Partition the matrix (B) into $t = M - S$ sub-matrices, (B_k) of order $S \times 2^{k-1}$ consisting of $q = 2^{k-1}$ vectors

$$(B_k) = (\boldsymbol{\beta}_{(q)}, \ldots, \boldsymbol{\beta}_{(2q-1)}) \text{ where } k = 1, 2, \ldots, t.$$ (9.30)

Correspondingly, we partition the random vector $\mathbf{c}_v$ into t sub-vectors $\mathbf{z}_k$, where

$$\mathbf{z}_k' = \left(C_{v,q}^{(M)}, C_{v,q+1}^{(M)}, \ldots, C_{v,2q-1}^{(M)} \right), q = 2^{k-1}, k = 1, 2, \ldots, n.$$ (9.31)

Define, for each $k = 1, \ldots, t, \mathbf{X}_k = (B_k)\mathbf{z}_k$. One can verify then that

$$\mathbf{n}_v = \sum_{k=1}^{n} \mathbf{X}_k$$ (9.32)

Let $\mathbf{W}_k' = (\mathbf{z}_0', \ldots, \mathbf{z}_k')$. Hence, under R.P.I., $P\{\mathbf{z}_k = W_{k-1}\} = P\{\mathbf{z}_k = -\mathbf{W}_{k-1}\} = \frac{1}{2}$.

Accordingly,

$$E\{\mathbf{X}_k\} = 0 \tag{9.33}$$

and due to orthogonality of the column vectors of $(C^{(M)})$,

$$E\{\mathbf{X}_{k_1}\mathbf{X}'_{k_2}\} = 0, \text{ for all } k_1 \neq k_2, \tag{9.34}$$

and

$$E\{\mathbf{X}_k\mathbf{X}'_k\} = (\mathbf{B}_k)(\mathbf{B}'_k). \tag{9.35}$$

Let

$$\mathbf{T}_m = \sum_{k=1}^{m} \mathbf{X}_k. \tag{9.36}$$

Due to the independence of $\mathbf{X}_k$, under R.P.I.,

$$E\{\mathbf{T}_m | \mathbf{T}_{m-1}\} = \mathbf{T}_{m-1}. \tag{9.37}$$

Thus, under R.P.I., $\{\mathbf{T}_m, m = 1, 2, \ldots\}$ is a martingale. The variance–covariance matrix of $\mathbf{T}_m$ is

$$E\{\mathbf{T}_m\mathbf{T}'_m\} = \sum_{k=1}^{m} (B_k)(B'_k). \tag{9.38}$$

As proved in Ehrenfeld and Zacks (1967), the following three conditions are necessary for the asymptotic normality of certain contrasts of interesting parameters:

1. For n sufficiently large, $\Xi_n = \sum_{k=1}^{n}(B_k)(B'_k)$ is positive definite, and can be factored to $\Xi_n = (Q_n)(Q'_n)$;

2. $\lim_{n \to \infty} \sup_{0 \leq i, j \leq S-1} \sup_{1 \leq k \leq t} \sum_{u=q}^{2q-1} |\beta_{j+u2^s}| \left(\sum_{k=1}^{n} \lambda_i^{(k)}\right)^{\frac{1}{2}} = 0$, where $\{\lambda_i^{(k)}, i = 0, \ldots, S - 1\}$ are the characteristic roots of $(B_k)(B'_k)$.

3. The randomization procedure is R.P.I.

As a corollary of this result, we obtain that under R.P.I. if

$$\lim_{n \to \infty} \sup_{1 \leq k \leq t} \frac{\sum_{u=q}^{2q-1} |\beta_{i+u2^s}|}{\left(\sum_{k=1}^{n} \sum_{u=q}^{2q-1} \beta_{i+u2^s}^2\right)^{\frac{1}{2}}} = 0, \tag{9.39}$$

then the asymptotic distribution of $\hat{\alpha}_i$, as $n \to \infty$ is $N\left(\alpha_i, \frac{\sigma^2}{S} + D_{i,n}^2\right)$ where

$$D_{i,n}^2 = \sum_{k=1}^{n} \sum_{u=q}^{2q-1} \beta_{i+u2^s}^2. \tag{9.40}$$

9.5   The ANOVA Scheme Under R.P.I

The hypothesis (9.28) can be tested on the basis of n independent estimates of $\boldsymbol{\alpha}$, obtained by R.P.I., by an ANOVA scheme in which the quadratic forms are formed, for each component of $\boldsymbol{\alpha}$, which are:

$$Q(\hat{\alpha}_{v_1,i}, \ldots, \hat{\alpha}_{v_n,i}) = S \sum_{j=1}^{n} (\hat{\alpha}_{v_j,i} - \hat{\alpha}_{.i})^2, \; i = 0, \ldots, S-1, \tag{9.41}$$

where

$$\hat{\alpha}_{.i} = \frac{1}{n} \sum_{j=1}^{n} \hat{\alpha}_{v_j,i}. \tag{9.42}$$

The conditional distribution of $\hat{\alpha}_{vi}$, given v, is $N\left(\alpha_i + n_{vi}, \frac{\sigma^2}{S}\right)$. Thus, under R.P.I., the distribution of $\hat{\alpha}_{vi}$ is the mixture $\left(\frac{1}{M}\right) \sum_{v=0}^{M-1} N\left(\alpha_i + n_{vi}, \frac{\sigma^2}{S}\right)$, for which $E\{\hat{\alpha}_{vi}\} = \alpha_i$ and $V\{\hat{\alpha}_{vi}\} = \frac{\sigma^2}{S} + \sum_{u=1}^{M-1} \beta_{i+u2^s}^2$. It follows that

$$E\{Q(\hat{\alpha}_{v_1,i}, \ldots, \hat{\alpha}_{v_n,i})\} = \sigma^2 + S \sum_{u=1}^{M-1} \beta_{i+u2^s}^2. \tag{9.43}$$

Similarly, let

$$Q^*(\hat{\alpha}_{.i}) = n S \hat{\alpha}_{.i}^2, \tag{9.44}$$

then

$$E\{Q^*(\hat{\alpha}_{.i})\} = \sigma^2 + S \sum_{u=1}^{M-1} \beta_{i+u2^s}^2 + n S \alpha_i^2. \tag{9.45}$$

This suggests that the F-like ratios

$$F^{(i)} = \frac{Q^*(\hat{\alpha}_{.i})}{Q(\hat{\alpha}_{v_1,i}, \ldots, \hat{\alpha}_{v_n,i})}, i = 0, \ldots, S - 1 \qquad (9.46)$$

are proper test statistics for testing the null hypotheses $H_0^{(i)} : \alpha_i = 0, \boldsymbol{\beta}$ arbitrary. The null distributions of these F-like ratios are quite complicated due to the influence of the nuisance parameters. If n is large, one could test these hypotheses simultaneously, with the Hotelling T^2 statistic. For this purpose, construct the $S \times S$ matrix

$$(V) = (A) \left[I^{(n)} - \frac{1}{n} \mathbf{J}^{(n)} \right] A', \qquad (9.47)$$

where $(A) = (\hat{\alpha}_1, \ldots, \hat{\alpha}_n)$ and the vector $\mathbf{Y} = \frac{1}{n} \sum_{i=1}^{n} \alpha_i$. Then,

$$T^2 = n(n-1) \mathbf{Y}'(V)^{-1} \mathbf{Y}. \qquad (9.48)$$

Under the above conditions of asymptotic normality, the critical level of this test statistic is

$$T^2_{1-\gamma} = \left[\frac{(n-1)S}{(n-2)} \right] F_{1-\gamma}[S, n - S]. \qquad (9.49)$$

A distribution free method of determining a critical level for T^2 is the *boot-strapping* method. From the set of LSE estimators $\{\hat{\alpha}_1, \ldots, \hat{\alpha}_n\}$, one draws a random sample, with replacement, of size n, designated as $\{\hat{\alpha}^*_1, \ldots, \hat{\alpha}^*_n\}$. With these values compute the statistic T^{2*}, according to formulas (9.47) and (9.48). Repeat this process a large number of times and compute the $(1 - \gamma)$ quantile of this boot-strapping set. Use this quantile as a critical level for the test.

9.6   Randomized Fractional Weighing Designs

The material in this section is based on the paper of Zacks (1966a,b). We start with a simple example to illustrate what are weighing designs. The weighing instrument is of the chemical balanced type. There are two pans, left and right. There are four objects to be weighted. We wish to have unbiased estimates of the four weight, with minimal total variance. Each weighing operation has a random error, ε, with $E\{\varepsilon\} = 0$, and $V\{\varepsilon\} = \sigma^2$. Only four weighings are allowed. If the four objects are weighted individually and independently in each one of the four operations, the total variance is $4\sigma^2$.

In Table 9.4 we present the optimal weighing design. If an object is put on the right pan we denote it by $+1$, if on the left pan we denote it by -1. WO denotes Weighing Operation; Ob denotes Object; Y denotes the weighing measurement.

Table 9.4 Optimal weighing
design for four objects

| | | Object | | | |
WO	Ob_1	Ob_2	Ob_3	Ob_4	Y
WO_1	1	-1	-1	1	Y_1
WO_2	1	1	-1	-1	Y_2
WO_3	1	-1	1	-1	Y_3
WO_4	1	1	1	1	Y_4

Let $\mathbf{Y}$ denote the vector of four measurements, and let $\mathbf{w}$ denote the vector of four weights. Notice that the matrix of coefficients in the table is the Hadamard matrix $(C^{(4)})$. Accordingly,

$$\mathbf{Y} = (C^{(4)})\mathbf{w} + \boldsymbol{\varepsilon},$$

and the best unbiased estimator of $\mathbf{w}$ is the LSE $\hat{\mathbf{w}} = \left(\frac{1}{4}\right)(C^{(4)})'\mathbf{Y}$. Moreover, $V\{\hat{\mathbf{w}}'1\} = \sigma^2$. This is the minimal possible risk. As shown by Mood (1946) such a weighing design based on a Hadamard matrix is optimal. We remark here that Hadamard matrix may not exit for an arbitrary p. We will assume here that $p = 2^s$ for which Hadamard matrix exists.

Consider a weighing problem of estimating a linear combination $\boldsymbol{\lambda}'\mathbf{w}$, where $\mathbf{w}$ is a vector of the true weights of p objects, when the number of weighing operations n is smaller than p. A design of such an operation is called *fractional weighing design*. If the fractional weighing design is not randomized, the design is singular, and one cannot estimate unbiasedly each one of the p weights. Some linear combinations can be estimated by a singular design: for example, the total weight. On the other hand, we show that any linear combination can be estimated unbiasedly under a proper randomization. Optimal randomization depends on the vector $\boldsymbol{\lambda}$.

9.6.1   Randomized Fractional Weighing Designs and Unbiased Estimation

Let (X) be a squared Hadamard matrix of order p. Let Y_i represent the observed value in the i-th weighing operation.

If p weighing operations are performed, the linear model is $\mathbf{Y} = (X)\mathbf{w} + \boldsymbol{\epsilon}$, where $E\{\boldsymbol{\varepsilon}\} = 0$, and $V[\boldsymbol{\varepsilon}] = \sigma^2 I$.

Let $\boldsymbol{\xi}$ be a probability vector of order p, and let $(\Xi) = diag\{\boldsymbol{\xi}\}$. (Ξ) is called a *randomization procedure matrix*. Let $(\mathbf{J})$ be a diagonal binary matrix of order p, with $J_{ii} = 1$, if the i-th row of (X) is chosen, and equal to 0 otherwise. $(\mathbf{J})$ is called an *allocation matrix*. It is evident that $E\{(\mathbf{J})\} = (\Xi)$. Let $(\mathbf{J})^{(j)}$ denote the allocation matrix for the j-th weighing operation ($j = 1, 2, \ldots, n$). Correspondingly, let $\mathbf{Y}^{(j)} = (\mathbf{J})^{(j)}(X\mathbf{w} + \boldsymbol{\varepsilon}^{(j)})$. This vector has one component different than zero, allocated at $(\mathbf{J})^{(j)}$.

Let $\mathbf{z}^{(n)} = \sum_{j=1}^{n} \mathbf{Y}^{(j)}$. Notice that at least $(n-p)$ components of $\mathbf{z}^{(n)}$ are zeros. A matrix $(\varXi)$ is non-singular if all diagonal elements are positive. The corresponding randomization procedure is called non-singular. We obtain the following result:

If ξ^* represents any non-singular randomization design, then

$$\hat{\mathbf{w}}^{(n)}(\xi^*) = (np)^{-1}(X)'(\varXi^*)^{-1}\mathbf{z}^{(n)} \tag{9.50}$$

is the unique linear unbiased estimator of $\mathbf{w}$, for every $n \geq 1$.

To prove the unbiasedness, substitute in (9.50) $E\{\mathbf{z}^{(n)}\} = n(\varXi^*)(X)\mathbf{w}$. Obviously, $\lambda'\hat{\mathbf{w}}^{(n)}(\xi^*)$ is an unbiased estimator of $\lambda'\mathbf{w}$. The class of unbiased estimators with non-singular randomization is incomplete with respect to the variance of the estimators. It is known that the minimal variance unbiased estimator of $1'\mathbf{w}$ corresponds to $\xi' = (1, 0'_{p-1})$, which is a singular design. In order to include also singular designs, we transform λ to $\alpha = p^{-1}(X)\lambda$. Some of the components of α might be 0. Suppose that r components of α, say $\alpha_{i_j}, j = 1, 2, \ldots, r; r \leq p$, are different from 0. Let $(\varXi(i_1, \ldots, i_r))$ be the corresponding randomization procedure matrix. The generalized inverse of $(\varXi)$ is

$$(\varXi)^- = \begin{pmatrix} \xi_1^- & \cdots & 0 \\ 0 & \cdots & 0 \\ 0 & \cdots & \xi_p^-, \end{pmatrix} \tag{9.51}$$

where $\xi_i^- = \xi_i^{-1}$, if $\xi_i \neq 0$, and $\xi_i^- = 0$, if $\xi_i = 0$. Thus, formula (9.50) is generalized to:

If only r components of α are different from zero, then

$$\lambda'\hat{\mathbf{w}}^{(n)}(\xi(i_1, \ldots, i_r)) = (np)^{-1}\lambda'(X)'(\varXi^-(i_1, \ldots, i_r))\mathbf{z}^{(n)} \tag{9.52}$$

is unbiased.

9.6.2 The Variance of an Unbiased Estimator

The variance–covariance matrix of the estimator in a non-singular design is

$$V\left[\hat{\mathbf{w}}^{(n)}(\xi^*)\right] = \frac{\sigma^2}{np^2} \sum_{i=1}^{p} \xi_i^{-1}\mathbf{X}_i\mathbf{X}_i^* + (np^2)^{-1} \sum_{i=1}^{p} \xi_i^{-1}\left(\mathbf{X}_i^*\mathbf{w}\right)^2 \mathbf{X}_i\mathbf{X}_i^* - n^{-1}\mathbf{w}\mathbf{w}'. \tag{9.53}$$

$\mathbf{X}_i'$ is the i-th row vector of (X). From this we obtain that the variance of the estimator (9.50) is

$$V\left\{\boldsymbol{\lambda}'\hat{\mathbf{w}}^{(n)}(\boldsymbol{\xi}^*)\right\} = \left(np^2\right)^{-1} \sum_{i=1}^{p} \xi_i^{-1} \left(\sigma^2 + \left(\mathbf{X}_i'\mathbf{w}\right)^2\right) \left(\boldsymbol{\lambda}'\mathbf{X}_i\right)^2 - n^{-1} \left(\boldsymbol{\lambda}'\mathbf{w}\right)^2.$$

(9.54)

If $\boldsymbol{\alpha}$ has r nonzero components, we apply the estimator (9.52), whose variance is

$$V\left\{\boldsymbol{\lambda}'\hat{\mathbf{w}}^{(n)}(\boldsymbol{\xi}^*)\right\} = \left(np^2\right)^{-1} \sum_{j=1}^{r} \xi_{i_j}^{-1} \left(\sigma^2 + \left(\mathbf{X}_{i_j}\mathbf{w}\right)^2\right) \left(\boldsymbol{\lambda}'\mathbf{X}_{i_j}\right)^2 - n^{-1} \left(\boldsymbol{\lambda}'\mathbf{w}\right)^2.$$

(9.55)

Notice that if

$$\begin{aligned}
\xi_i^* &= \frac{\xi_{i_j}}{\sum_{j=1}^{r} \xi_{i_j}}, && \text{if } i = i_j, j = 1, \ldots, r, \\
&= 0, && \text{otherwise,}
\end{aligned}$$

(9.56)

then $V\{\boldsymbol{\lambda}'\hat{\mathbf{w}}^{(n)}(\boldsymbol{\xi}^*(i_1, \ldots, i_r))\} < V\{\boldsymbol{\lambda}'\hat{\mathbf{w}}^{(n)}(\boldsymbol{\xi}^*)\}$. That is, if $r < p$, the variance of the unbiased estimator under $\boldsymbol{\xi}^*(i_1, \ldots, i_r)$ is strictly smaller than that of the unbiased estimator under $\boldsymbol{\xi}^*$. This shows us how to improve the design in certain cases.

Chapter 10
Sequential Search of an Optimal Dosage

New drugs are subjected to clinical trials before being approved for usage. These clinical trials consist of several phases. In phase I the objective is to establish the maximal tolerated dose (MTD). The assumption is that the effectiveness of a drug is a monotone increasing function of the dosage applied. On the other hand, there are always side effects due to the toxicity of the drugs. The level of toxicity tolerated by different subjects varies at different dosages. There is a tolerance distribution in a population of subjects (patients), which is a function of the dosage. There are cases in which the level of toxicity can be measured on a linear scale and is represented by a continuous random variable $Y(x)$, where x is the dosage. The first part of the present chapter deals with such a model, following the paper of Eichhorn and Zacks (1973). The second part of the chapter deals with cases in which the level of toxicity is represented by a binary variable $J(x)$, where $J(x) = 1$, if the level of toxicity is dangerous to the life of the patient, and $J(x) = 0$, otherwise. A dangerous dose is called "lethal doese, LTD." An MTD is a maximal dosage such that the probability of an lethal doese, LTD is smaller than a prescribed threshold γ (in many cases $\gamma = 0.3$). A sequential search for an MTD applies the drug to individuals, one by one, starting from a "safe" dosage, and increasing or decreasing the dosage according to the observed toxicity levels.

10.1 The Linear Models

We consider the following regression model. The toxicity level is negligible for all $0 < x < x_0$. For $x \geq x_0$, $E\{Y(x)\} = b(x - x_0)$. The value of x_0 is known. We have two models for the conditional distribution of $Y(x)$ given x.

Model 1: $Y(x)|x \sim N(b(x - x_0), \sigma^2(x - x_0)^2)$, σ^2 is known and b is unknown, $0 < b < \infty$.

Model 2: $Y(x)|x \sim N(b(x - x_0), \sigma^2)$.

© Springer Nature Switzerland AG 2020
S. Zacks, *The Career of a Research Statistician*, Statistics for Industry,
Technology, and Engineering, https://doi.org/10.1007/978-3-030-39434-9_10

In Model 2, the variance σ^2 is known, and b is unknown. Let η designate a dangerous toxicity level. It is desired that

$$P\{Y(x) \le \eta | x\} \ge \gamma, \quad \text{for all } x < \xi_\gamma, \tag{10.1}$$

where

$$x_\gamma = x_0 + \frac{\eta}{(b + \sigma Z_\gamma)}, \quad \text{for Model 1,} \tag{10.2}$$

$$= x_0 + \frac{\eta - Z_\gamma \sigma}{b}, \quad \text{for Model 2} \tag{10.3}$$

$$Z_\gamma = \Phi^{-1}(\gamma).$$

We seek a sequence of dosages $x_1, x_2, \ldots$ satisfying the following conditions:

1. For each $n = 1, 2, \ldots$, $P_{b,\sigma}\{x_n \le \xi_\gamma\} \ge 1 - \alpha$, for all $0 < b, \sigma < \infty$.
2. $\lim_{n \to} x_n = \xi_\gamma$ in probability.

A sequence satisfying (1) is called *feasible*, and a sequence satisfying (2) is called *consistent*.

10.1.1   A Search Procedure for Model 1

Given the values of $x_1, \ldots, x_n$ and the associated values of $Y_1, \ldots, Y_n$ we define $U_i = \frac{Y_i}{(x_i - x_o)}, i = 1, \ldots, n$, and

$$\bar{U}_n = \frac{1}{n} \sum_{i=1}^{n} U_i. \tag{10.4}$$

Notice that $U_i | x_i \sim N(b, \sigma^2)$. Hence, according to (10.2) if b is unknown an estimator of the MTD is

$$\hat{x}_n = x_0 + \frac{\eta}{\bar{U}_n + \sigma Z_\gamma}.$$

Since we wish that the estimator of the MTD will be smaller than ξ_γ with probability $(1 - \alpha)$, we set the next dosage to be

$$x_{n+1} = x_0 + \frac{\eta}{\bar{U}_n^+ + \sigma \left(Z_\gamma + Z_{1-\alpha}/\sqrt{n}\right)}, n \ge 1. \tag{10.5}$$

The first dosage x_1 should be greater than x_0, but otherwise an arbitrary constant, say x^*, for which we know that the toxicity there is negligible.

From (10.2) we obtain that $\frac{\eta}{\xi_\gamma - x_0} = b + \sigma Z_\gamma$. This implies that

$$P\left\{x_0 + \frac{\eta}{\bar{U}_n^+ + \sigma\left(Z_\gamma + Z_{1-\alpha}/\sqrt{n}\right)} \leq \xi_\gamma\right\}$$

$$= P\left\{\bar{U}_n^+ + \sigma\left(\frac{Z_{1-\alpha}}{\sqrt{n}}\right) \geq b\right\} \geq P\left\{\bar{U}_n + \sigma\left(\frac{Z_{1-\alpha}}{\sqrt{n}}\right) \geq b\right\} \qquad (10.6)$$

$$= 1 - \alpha.$$

Thus, $P\{x_n \leq \xi_\gamma\} \geq 1 - \alpha$, for all $n \geq 1$, and the sequence $\{x_n, n \geq 1\}$ is feasible.

Also, by the strong law of large numbers, $\bar{U}_n \to b$, a.s., and $b > 0$. Thus, the sequence of dosages is strongly consistent. It is also proved in Eichhorn and Zacks (1973) that among the feasible sequences, the sequence given in (10.5) is uniformly most accurate in the sense of Lehmann (1959, pp. 78–81).

If σ^2 is unknown, we can estimate it by the sample variance of the U's, namely $S_n^2 = \frac{1}{n-1} \sum_{i=1}^{n} (U_i - \bar{U}_n)^2$. We can use then the sequence of dosages

$$x_{n+1} = x_0 + \frac{\eta}{\bar{U}_n + S_n(t_\zeta[n-1]\gamma + t_{1-\alpha}[n-1]/\sqrt{n}}, n \geq 2.$$

For $n = 0, 1$ we can use two arbitrary values close to x_0.

10.1.2   A Search Procedure for Model 2

In Model 2, as in Model 1, $U_i = \frac{Y(x_i)}{(x_i - x_0)}$, and $E\{U_i\} = b$. However, $V\{U_i | x_i\} = \frac{\sigma^2}{(x_i - x_0)^2}$. Accordingly, the distributions of $U_i, i = 1, 2, \ldots$ are not normal, and these variables are not independent when the dosages x_i are functions of the previous observations. Notice that $S_n = \sum_{i=1}^{n} U_i$ are sub-martingales, since $b > 0$. Moreover, $COV(Y_i, Y_j) = 0$, for all $i < j$. Accordingly,

$$V\{\bar{U}_n\} = \frac{\sigma^2}{n^2} \sum_{i=1}^{n} E\left\{\frac{1}{(x_i - x_0)^2}\right\}. \qquad (10.7)$$

We require that $x_i \geq x_0^*$. Hence,

$$V\{\bar{U}_n\} \leq \frac{\sigma^2}{n}(x_0^* - x_0)^{-2}. \qquad (10.8)$$

This leads us to suggest the following sequence of dosages under Model 2:

$$x_1 = x_0^* \tag{10.9}$$

$$x_{n+1} = \max \left\{ x_0^*, x_0 + \frac{\eta - Z_\gamma \sigma}{\bar{U}_n + \dfrac{d_\alpha \sigma}{\sqrt{n}}} \right\}, n \geq 1. \tag{10.10}$$

Here

$$d_\alpha = \frac{1}{\alpha^{\frac{1}{2}} (x_0^* - x_0)}. \tag{10.11}$$

This is a conservative approach, guaranteeing (by Chebychev's inequality) that

$$P \left\{ |U_n - b| < d_\alpha \left(\frac{\sigma}{\sqrt{n}} \right) \right\} \geq 1 - \alpha. \tag{10.12}$$

10.1.3 Bayes Procedures

In the Bayesian framework we start with a prior distribution for the unknown parameter b. Let $H(b)$ denote the prior c.d.f., and $H(b|\mathbf{Y}_n, \mathbf{X}_n)$ denote the posterior c.d.f., given the results of the first n observations. We require that the dosages will be determined so that

$$P\{x_{n+1} \leq \xi_\gamma | \mathbf{Y}_n, \mathbf{X}_n\} \geq 1 - \alpha, \text{ for every } n = 1, 2, \ldots \tag{10.13}$$

This requirement is weaker than the feasibility requirement for the non-Bayesian procedures. This Bayesian requirement is called *Bayes-feasible*. We provide below explicit formulas for the case where the prior distribution is normal $N(\beta, V_0)$. One could choose the prior parameters so that the prior probability that $b < 0$ is negligible. In the above notation, $\mathbf{X}_n = (X_1, \ldots, X_n)$, where $X_i = x_i - x_0$. As before, $U_i = \frac{Y_i}{X_i}$. Also

$$\tau_n^2 = \sigma^2 X_n^2 \text{ for Model 1} \tag{10.14}$$

$$= \sigma^2 \text{ for Model 2 .} \tag{10.15}$$

Given $(\mathbf{Y}_n, \mathbf{X}_n)$ the posterior distribution of b is $N(\beta_n, V_n)$, where

$$\beta_n = \beta_{n-1} + \frac{(Y_n - \beta_{n-1} X_n) X_n V_{n-1}}{\tau_n^2 + X_n^2 V_{n-1}}, n = 1, 2, \ldots \tag{10.16}$$

and

$$V_n = \frac{V_{n-1}\tau_n^2}{\tau_n^2 + X_n^2 V_{n-1}}. \tag{10.17}$$

This recursive equation can be simplified for Model 1 to obtain,

$$V_n = \frac{\sigma^2}{n + \dfrac{\sigma^2}{V_0}}; \tag{10.18}$$

$$\beta_n = \frac{\bar{U}_n}{1 + \dfrac{\sigma^2}{nV_0}} + \frac{\beta\left(1 + \dfrac{\sigma^2}{V_0}\right)}{n + \dfrac{\sigma^2}{V_0}}. \tag{10.19}$$

Thus, under Model 1, $V_n \to 0$, as $n \to \infty$, and $\beta_n \to b$, a.s.
The Bayes procedure specifies the following sequence of dosages

$$x_{n+1} = \max\{x_n^*, \xi_{n,\gamma}\}, \tag{10.20}$$

where

$$\xi_{n,\gamma} = x_0 + \frac{\eta}{\beta_n + \sigma Z_\gamma + Z_{1-\alpha}\sqrt{V_n}} \quad \text{for Model I} \tag{10.21}$$

$$= x_0 + \frac{\eta - Z_\gamma \sigma}{\beta_n + Z_{1-\alpha}\sqrt{V_n}} \quad \text{for Model 2}. \tag{10.22}$$

10.1.4 Numerical Comparison

In the present section we compare the above procedures numerically.

We start with an initial dose x_1, then simulate the toxicity value Y_1, from the normal distribution, and so on. The parameters of this simulation are: $b = 3$, $x_0 = 0$, $x_0^* = 1$, $\eta = 10$, $\sigma = 1$, $\alpha = 0.05$, $\gamma = 0.99$. The values of the optimal dosages are:

$$\xi_\gamma = 1.878 \text{ for Model 1}$$
$$= 2.558 \text{ for Model 2}.$$

The Bayesian prior distribution of b is $N(2.86, 0.25)$. The simulated dosages are given in Table 10.1.

Table 10.1 Simulated
dosages, non-Bayes (N.B.),
Bayes (B)

	Model 1		Model 2	
n	N.B.	B	N.B.	B
1	1.1461	1.5878	1.0000	2.0460
2	1.2675	1.5701	1.1064	2.0513
3	1.3485	1.5737	1.2211	2.0765
4	1.4275	1.5963	1.3313	2.1221
5	1.4264	1.5778	1.3693	2.1101
6	1.4955	1.6122	1.4566	2.1652
7	1.5373	1.6328	1.5206	2.2010
8	1.5559	1.6395	1.5617	2.2171
9	1.6235	1.6847	1.6420	2.2827
10	1.6194	1.6779	1.6600	2.2787
15	1.6984	1.7296	1.8068	2.3618
20	1.7456	1.7646	1.9027	2.4142
25	1.7298	1.7481	1.9316	2.3986
30	1.7742	1.7857	2.0021	2.4472
35	1.7678	1.7787	2.0238	2.4412
40	1.7639	1.7741	2.0426	2.4377
45	1.8026	1.8091	2.0946	2.4791
50	1.8083	1.8140	2.1163	2.4855

It seems that in Model 1, both the non-Bayesian and the Bayesian procedures converge to the optimal dosage in the same pace. In Model 2 the Bayesian procedure is better due to the over-pessimistic non-Bayes procedure. See also Eichhorn (1974) and Eichhorn and Zacks (1981).

10.2   Categorical Model for Dosage Toxicity

In contrast to the linear models of the previous section, in categorical models the response variable $Y(x)$ is discrete. In the following we assume that the response variable is binary, i.e., $Y(x) = 1$ if the response represents a LDT, and $Y(x) = 0$ otherwise. The relationship between the response variable and the dosage x is

$$P\{Y(x) = 1|x\} = F(\beta_0 + \beta_1 x), \tag{10.23}$$

where $\beta_1 > 0$ and $F(x)$ is a standard distribution function (c.d.f.) called a *tolerance distribution*. The MTD is a dosage x_γ such that $F(\beta_0 + \beta_1 x_\gamma) = \gamma$. Or,

$$x_\gamma = \frac{F^{-1}(\gamma) - \beta_0}{\beta_1}. \tag{10.24}$$

The logistic tolerance distribution is often applied in the biostatistics literature, i.e.,

$$F(\beta_0 + \beta_1 x) = \frac{\exp\{\beta_0 + \beta_1 x\}}{1 + \exp\{\beta_0 + \beta_1 x\}}. \tag{10.25}$$

Suppose it is known that all dosages should be in the interval $[x^*, x^{**})$, and that the probability of LDT at x^* is p_1. In this case we can write the logistic tolerance distribution as a function

$$Q(x) = \frac{p_1 \exp\{\beta(x - x^*)\}}{(1 - p_1 + p_1 \exp\{\beta(x - x^*)\})}, \quad \text{for } \beta > 0. \tag{10.26}$$

Notice that in this special case the MTD is

$$x_\gamma = x^* + \frac{\log\left(\frac{\gamma}{1-\gamma}\right)\log\left(\frac{p_1}{1-p_1}\right)}{\beta}. \tag{10.27}$$

Since β is unknown, the problem is to design a sequence of dosages which with high probability will be below the MTD and will converge to the MTD. This will be discussed in the following section.

10.2.1   Bayesian Adaptive Search for the MTD

As before we consider a sequential procedure. Each patient is given a dosage which is determined by the previous results. Let $D^{(k)} = \{(x_j, Y_j) : j = 1, \ldots, k\}$ denote the data observed after the first k trials. The likelihood function of β given $D^{(k)}$ is

$$L_k(\beta, D^{(k)}) = \exp\left\{ -\sum_{j=1}^k \left[\log\left(1 + \frac{p_1}{1 - p_1}\right) \right.\right.$$
$$\left.\left. \times \exp\{\beta(x_j - x^*)\} - \beta Y_j(x_j - x^*) \right] \right\}. \tag{10.28}$$

With this likelihood function and a prior distribution of β we obtain the posterior distribution of β. The $(1 - \alpha)$-quantile of the posterior distribution is substituted in (10.27) for β. This procedure yields a sequence of dosages called *Bayes feasible*. More formally, let $h(\beta)$ denote a prior density for β. According to Bayes theorem the posterior density of β given $D^{(k)}$ is

$$h(\beta|D^{(k)}) = \frac{h(\beta)L_k(\beta, D^{(k)})}{\int_0^\infty h(t)L_k(t, D^{(k)})dt}. \tag{10.29}$$

We denote by $H^{-1}(1-\alpha|D^{(k)})$ the $(1-\alpha)$-quantile of the posterior distribution of β. Thus, the $(k+1)$ dosage given $D^{(k)}$ is

$$x_\gamma = x^* + \frac{\log\left(\frac{\gamma}{1-\gamma}\right)\log\left(\frac{p_1}{1-p_1}\right)}{H^{-1}(1-\alpha|D)}. \tag{10.30}$$

This procedure of Bayesian determination of dosages was called *Escalation with Over-dose Control* (EWOC), see Rogatko, Babb, and Zacks (1998). See the article of Zacks et al. (1998). We illustrate below the EWOC procedure by a numerical simulation. For this purpose we adopt a logistic model used by Durham and Flournoy (1995), which is

$$Q(x) = \frac{\exp\{-3.569 + 0.549x\}}{1 + \exp\{-3.569 + 0.549x\}}. \tag{10.31}$$

A few values of this distribution are given in Table 10.2.

For $\gamma = 0.33$ the MTD is 5.24. Here $x^* = 2.5$ and $x^{**} = 7.25$. The prior distribution for β is the Gamma distribution with scale parameter 1 and shape parameter $\nu = 1, \frac{1}{2}, \frac{1}{3}, \frac{1}{4}$. The values of $Y(x)$ are simulated from the Binomial distribution $B(1, Q(x))$. For the posterior quantile we use the posterior mean plus three times the posterior standard deviation. This will give $\alpha < 0.1$. The computations are performed according to the R-function "bayesim" given in the Appendix.

We see in Table 10.3 that the results are sensitive to the choice of the shape parameter. For $\nu = \frac{1}{4}$ there are no dosage above the MTD. For $\nu = \frac{1}{2}$ the sequence climbed faster to the neighborhood of the MTD, but there are nine cases with dosages above the MTD.

We conclude the chapter with the following comment. In the literature there are several papers in which the Robbins Monroe stochastic approximation is applied for approximating the MTD. See for example the paper of Anbar (1984). In these papers the authors show that the estimator of the MTD is asymptotically normally distributed around the MTD. In these large samples there are approximately 50% dosages generated by these sequences will be above the MTD. The EWOC method

Table 10.2 The distribution of $Q(x)$

x	2.5	4.5	5.24	5.76	6.5	7.25
$Q(x)$	0.10	0.25	0.33	0.40	0.50	0.60

Table 10.3 Sequences of dosages according to EWOC

n	$v = 1$	$v = \frac{1}{2}$	$v = \frac{1}{3}$	$v = \frac{1}{4}$
1	2.500	2.500	2.500	2.500
2	3.802	3.243	3.064	2.956
3	4.296	3.420	3.171	3.012
4	4.932	3.627	3.230	3.038
5	5.745	3.817	3.371	3.079
6	6.815	4.049	3.408	3.119
7	5.102	4.384	3.492	3.162
8	4.459	4.725	3.598	3.229
9	4.574	3.979	3.305	3.288
10	4.840	4.067	3.387	3.352
11	5.011	3.634	3.453	3.428
12	5.309	3.753	3.483	3.505
13	5.625	3.815	3.604	3.555
14	4.971	3.891	3.619	3.726
15	5.011	4.001	3.781	3.742
16	4.742	4.146	3.413	3.825
17	4.847	4.291	3.516	3.874
18	4.506	4.432	3.515	4.044
19	4.349	4.616	3.580	4.171
20	4.418	4.828	3.688	4.209
21	4.485	5.0956	3.786	4.329
22	4.375	5.356	3.860	3.868
23	4.367	5.658	3.522	3.981
24	4.422	6.055	3.626	4.038
25	4.479	6.531	3.647	4.191
26	4.559	7.005	3.705	4.190
27	4.393	7.250	3.476	4.347
28	4.320	6.060	3.522	4.378
29	4.359	5.333	3.608	4.477
30	4.406	5.505	3.648	4.685

tries to avoid such results, and aims that the sequence of dosages will be Bayes feasible. See for example Fig. 10.1 in which the EWOC procedure is illustrated with a sample of size 300. Remember however that in Phase I clinical trials the sample is generally small.

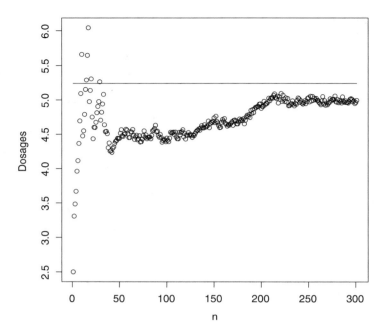

Fig. 10.1 Progression of dosages in EWOC, MTD= 5.24

Chapter 11
Contributions to Applied Probability

In the present chapter, I present mainly distributions of stopping times, which are the first epochs at which compound processes cut given boundaries. These distributions were developed for exact computations of required functionals in various applications in inventory theory, queuing theory, and more. The material discussed here is restricted to the fundamental theory, with numerical illustrations. The reader is referred to the recent book of Zacks (2017). R-functions can be found in the Appendix.

11.1 The Compound Poisson Process

Consider the problem of tracking the income from sale of a given product at a store. Customers arrive at the store independently of each other, and the amount spent is independent of the arrival time. The quantities bought by customers are independent random variable. The income process is a jump process which, under the conditions specified below, is called a *compound Poisson process*.

Another example could be the accumulations of claims from accidents at an insurance company. The model of compound Poisson process is used also in queuing theory to describe the random amount of work required from a server, in a one server station. It is assumed that customers arrive at the station one by one at random times, following a Poisson process, and the required amounts of work are i.i.d. random variables, independent of the arrival times.

The following is the fundamental theory.

Let $\{N(t), t \geq 0\}$ be a Poisson process. Such a process was defined in Chap. 8, as a process of independent increments, where $N(t)$ is a random variable having a Poisson distribution, with mean λt. Moreover, $N(t)$ is a counting process representing the number of arrivals in the time interval $(0, t]$.

© Springer Nature Switzerland AG 2020
S. Zacks, *The Career of a Research Statistician*, Statistics for Industry,
Technology, and Engineering, https://doi.org/10.1007/978-3-030-39434-9_11

Let $\{X_1, X_2, \ldots\}$ be a sequence of i.i.d. random variables, independent of $\{N(t), t > 0\}$, which describe the "rewards" associated with the random arrivals. Thus, the value of the process at time t is

$$Y(t) = \sum_{n=1}^{N(t)} X_n. \tag{11.1}$$

Obviously $Y(t) = 0$ as long as $N(t) = 0$. Let F denote the distribution of X, and f the density of F. We assume in the present section that $X \geq 0$. Since the $X's$ are i.i.d., $P\{Y(t) \leq y | N(t) = n\} = F^{(n)}(y)$, where $F^{(n)}$ denotes the n-fold convolution of F defined recursively as

$$F^{(0)}(y) = 1, \text{ for all } y \geq 0,$$

$$F^{(n)}(y) = \int_0^y f(x) F^{(n-1)}(y - x) dx, \text{ for } n \geq 1.$$

It follows that the c.d.f. of $Y(t)$ is

$$H(y, t) = \sum_{n=0}^{\infty} p(n; \lambda t) F^{(n)}(y). \tag{11.2}$$

The corresponding density of H is

$$h(y, t) = \sum_{n=1}^{\infty} p(n, \lambda t) f^{(n)}(y). \tag{11.3}$$

Due to independence, $E\{Y(T)\} = E\{E\{Y(t)|N(t)\}\} = E\{N(t)\xi\} = \xi \lambda t$, where $\xi = E\{X\}$. Similarly, $V\{(Y(t)\} = (\sigma^2 + \xi^2)\lambda t$, where $\sigma^2 = V\{X\}$. Let $\psi(\theta)$ denote the moment generating function of X. The moment generating function of $Y(t)$ is

$$M(\theta, t) = \exp\{-\lambda t(1 - \psi(\theta))\}. \tag{11.4}$$

Equation (11.4) is valid for all θ in the domain of convergence of $\psi(\theta)$. The c.d.f. of $Y(t)$ can be easily determined when F is Exponential with mean $\beta = \frac{1}{\mu}$. In this case $F^{(n)}$ is the Gamma distribution with scale parameter β and shape parameter $\nu = n$. According to the Gamma–Poisson relationship, $F^{(n)}(y) = 1 - P(n-1, \mu y)$. It follows that in this case

$$H(y, t) = 1 - \sum_{n=0}^{\infty} p(n, \lambda t) P(n-1, \mu y) = \sum_{n=0}^{\infty} p(n, \mu y) P(n, \lambda t).$$

In Table 11.1 we present the function $H(y, t)$.

Table 11.1 Values of
$H(y, t)$ for $\lambda = 1$ and
$\mu = 1, 2$ $t = 1, 2$

λ	μ	y	$H(y, 1)$	$H(y, 2)$
1	1	0.05	0.38604	0.14887
		0.07	0.39318	0.15428
		0.10	0.40379	0.16239
		0.50	0.53013	0.26901
		1.00	0.65425	0.39430
		2.00	0.81742	0.60350
		2.50	0.86870	0.68537
		3.00	0.90614	0.75301
		4.00	0.95277	0.85194
1	2	0.05	0.40376	0.16239
		0.07	0.41761	0.17319
		0.10	0.43786	0.18936
		0.50	0.65426	0.39430
		1.00	0.81742	0.60350
		2.00	0.95278	0.85194
		2.50	0.97665	0.91393
		3.00	0.98863	0.95223
		4.00	0.99740	0.98528

11.2   First Crossing Linear Boundaries

In the present section we derive the exact distribution of the first time at which a compound Poisson process with positive jumps crosses a linear boundary. We also assume that the distribution F is absolutely continuous. An example of application can be given from inventory control. Suppose there is a commodity which is perishable after t_0 time units (days). The demand process is compound Poisson with known parameters. The question is: How many units to store, so that the demand can be satisfied and not too many units are discarded at the end. This example will be discussed later. Additional reading about boundary crossing of compound Poisson process with positive jumps can be found in Perry, Stadje, and Zacks (1999, 2002), Zacks (1991), and Zacks, Perry, Bshuty, and Bar-Lev (1999).

11.2.1   Nonincreasing Linear Boundaries

We start with a horizontal boundary, at level β. The stopping time is the first crossing of the level β, i.e.,

$$T_0(\beta) = \inf\{t > 0 : Y(1) \geq \beta\}. \tag{11.5}$$

Since $Y(t)$ is nondecreasing

$$P\{T_0(\beta) > t\} = P\{Y(s) < \beta, \text{ for all } s \leq t\} = H(\beta, t). \tag{11.6}$$

The density of $T_0(\beta)$ is

$$\psi_0(t) = -\frac{d}{dt} H(\beta, t) = \lambda \sum_{n=0}^{\infty} \left[F^{(n)}(\beta) - F^{(n+1)}(\beta) \right] p(n, \lambda t). \tag{11.7}$$

Furthermore, since $F^{(n+1)}(\beta) = \int_0^\beta f(y) F^{(n)}(\beta - y) dy$, we obtain the alternative formula

$$\psi_0(t) = \lambda e^{-\lambda t}(1 - F(\beta)) + \lambda \int_0^\beta h(y, t)(1 - F(\beta - y)) dy. \tag{11.8}$$

This formula is very useful when analytic expressions for the convolutions of F are not available. The formula is known as a *level crossing* formula.

Consider now the decreasing linear boundary $B(t) = \beta - t, 0 < t \leq \beta$. The stopping time is $T_1(\beta) = \inf\{t > 0 : Y(t) \geq \beta - t\}$. With similar arguments we get

$$P\{T_1(\beta) > t\} = H(\beta - t, t). \tag{11.9}$$

In the present case there are two modes of crossing the boundary. Mode I is crossing horizontally, i.e., $Y(t) = \beta - t$. Mode II is crossing by jumping above the boundary, i.e., $Y(t) > \beta - t$. Accordingly, the density of the stopping time has two components

$$\psi_1^{(1)}(t) = h(\beta - t, t), \tag{11.10}$$

and

$$\psi_1^{(2)}(t) = \lambda \sum_{n=0}^{\infty} p(n, \lambda t) \left[F^{(n)}(\beta - t) - F^{(n+1)}(\beta - t) \right]. \tag{11.11}$$

The density is the sum of these two components, or

$$\psi_1(t) = \lambda e^{-\lambda t}(1 - F(\beta - t)) + \lambda \int_0^{\beta - t} h(y, t)(1 - F(\beta - t - y)) dy + h(\beta - t, t).$$

The k-th moment of $T_1(\beta)$ is

$$E\left\{T_1(\beta)^k\right\} = k \int_0^\beta s^{k-1} H(\beta - s, s) ds. \tag{11.12}$$

Table 11.2 The survival
distribution of $T_1(10)$

t	$H(10 - t, t)$
1	0.9988
2	0.9853
3	0.9278
4	0.7876
5	0.5639
6	0.3180
7	0.1302
8	0.0341
9	0.0043
10	$e(-10)$

If X has an exponential distribution, with mean $1/\mu$, the density of $T_1(\beta)$ is

$$\psi_1(t) = \mu \sum_{n=0}^{\infty} p(n, \lambda t) p(n - 1, \mu(\beta - t)) \left(1 + \frac{\lambda(\beta - t)}{n} \right). \qquad (11.13)$$

The k-th moment is

$$E\{T_1^k(\beta)\} = k \sum_{n=0}^{\infty} \int_0^{\beta} s^{k-1} p(n, \mu(\beta - s)) P(n, \lambda s) ds. \qquad (11.14)$$

In Table 11.2 we present the survival function $P\{T_1(\beta) > t\}$ for the exponential case, with $\mu = 1$, $\lambda = 1$, $\beta = 10$.
In addition $E\{T_1(10)\} = 5.25$.

11.2.2   *Increasing Linear Boundaries*

11.2.2.1   Case I: Negative Intercept

An important application of the following theory is in computing the distribution of the busy period in an M/G/1 queue. This is a station with one server, to which customers arrive according to a Poisson process, and the service length of each customer is a random variable with a general distribution. The problem is to characterize the length of the period in which the server is continuously busy.
   The boundary is $B(t) = -\beta + t$, $\beta > 0$. The corresponding stopping time is

$$T_L(\beta) = \inf\{t > 0 : Y(t) = -\beta + t\}, t \geq \beta. \qquad (11.15)$$

Since the sample path of $Y(t)$ is nondecreasing, $P\{T_L(\beta) < \infty\} = 1$ only if $E\{X\} < 1/\lambda$. Furthermore, $P\{T_L(\beta) = \beta\} = \exp\{-\lambda\beta\}$.

Table 11.3 Distribution of $T_L(\beta)$

t	$\Psi_L(t; 5)$
5	0.006738
6	0.223079
7	0.489100
8	0.689948
9	0.819348
10	0.897044
11	0.942051
12	0.967614
13	0.981972
14	0.989985
15	0.996723
Inf	1.000000

As proved in Zacks (2017), the density of $T_L(\beta)$ is

$$\psi_L(t; \beta) = \left(\frac{\beta}{t}\right) h(t - \beta, t), t > \beta. \tag{11.16}$$

The first two moments are

$$E\{T_L(\beta)\} = \frac{\beta}{1 - \rho}, \tag{11.17}$$

and

$$E\{T_L^2(\beta)\} = \frac{\beta^2}{(1 - \rho)^2} + \frac{\beta\lambda\mu_2}{(1 - \rho)^3}, \tag{11.18}$$

where $\rho = E\{X\}\lambda$, and $\mu_2 = E\{X^2\}$. In Table 11.3 we present the c.d.f. of $T_L(\beta)$ which is $\Psi_L(t; \beta) = e^{-\lambda\beta} + \beta \int_\beta^t (\frac{1}{s}) h(s - \beta, s) ds$, for $\lambda = 1, \mu = 3, \beta = 5$.

The mean and variance of this distribution are: $E\{T_L(\beta)\} = 7.5$ and $V\{T_L(\beta)\} = 2.6337$.

Application of this result in queuing theory will be shown in Sect. 11.6.

11.2.2.2 Case II: $B(t) = \beta + t, \beta > 0$

Applications will be shown in insurance (actuarial science), when the total demand for claims surpasses the income from premiums. The stopping time is $T_U = \inf\{t > 0 : Y(t) \geq B(t)\}$. In this case $P\{T_U < \infty\} = 1$ only if $\lambda E\{X\} > 1$.

We start with the case of $\beta = 0$.

Introduce the defective density of $Y(t)$, namely

$$g_0(y, t) = \frac{d}{dy} P\{Y(t) \leq y, Y(s) < s \text{ for all } 0 < s \leq t\}. \tag{11.19}$$

Obviously, $g_0(y, t) = 0$ for all $y \geq t$. For all $0 < y < t$,

$$g_0(y, t) = h(y, t) - \int_0^t h(x, x)g_0(y - x, t - x)dx - h(y, y)e^{-\lambda(t-y)}. \quad (11.20)$$

The function $g_0(y, t)$ is the solution of this integral equation. Stadje and Zacks (2003) proved that

$$g_0(y, t) = \left(\frac{t - y}{t}\right) h(y, t), 0 < y < t. \quad (11.21)$$

It follows that in this case of $\beta = 0$,

$$P\{T_U > t\} = e^{-t\lambda} + \int_0^t g_0(y, t)dy. \quad (11.22)$$

The corresponding density of T_U can be obtained from the level crossing equation

$$\psi_U(t) = \lambda e^{-\lambda t}(1 - F(t)) + \lambda \int_0^t g_0(y, t)(1 - F(t - y))dy. \quad (11.23)$$

In Table 11.4 we present the survival function of T_U for $\lambda = 1.2$, $\mu = 1$ for the case of $F(t) = 1 - \exp\{-\mu t\}$.

When $\beta > 0$, the stopping time is $T_U(\beta) = \inf\{t > 0 : Y(t) \geq \beta + t\}$. The defective density of $Y(t)$ is in this case

$$g_\beta(y, t) = I\{y \leq \beta\}h(y, t) + I\{\beta < y < \beta = t\}[h(y, t)$$

$$- \int_\beta^{\beta+t} h(x, x - \beta)g_0((y - x, t - x)dx - h(y, y + t) \quad (11.24)$$

$$\times \exp\{-\lambda(t + \beta - y)\}].$$

Table 11.4 Survival function of T_U

t	$P\{T_U > t\}$
0	1.0000
0.25	0.7667
0.5	0.6223
1.0	0.4588
1.5	0.3698
2	0.3133
3	0.2441
5	0.1734
10	0.1027

Table 11.5 Survival
distribution of $T_U(\beta)$, for
$\beta = 3$, F exponential with
$\mu = 1$

t	$\lambda = 1.2$	$\lambda = 2$
0.5	0.96134	0.92136
1	0.90776	0.81047
2	0.78511	0.60151
3	0.68069	0.45529
4	0.60520	0.35466
5	0.55278	0.28174
6	0.51600	0.22632
7	0.48924	0.18296
10	0.43851	0.09857
20	0.34894	0.01386

The corresponding density of $T_U(\beta)$ is

$$\psi_U(t; \beta) = \lambda e^{-\lambda t}(1 - F(\beta + t)) + \lambda \int_0^{\beta+t} g_\beta(x, t)(1 - F(\beta + t - x))dx. \qquad (11.25)$$

The survival distribution is $P\{T_U(\beta) > t\} = H(\beta, t) + \int_\beta^{\beta+t} g_\beta(y, t)dy$ (Table 11.5). See also the papers of Zacks (1991) and Zacks, Perry, Bshuty, and Bar-Lev (1999).

11.3 Fraction of Time in a Mode

The material in this section is based on the paper of Zacks (2012). Consider a process $\{U_1, D_1, U_2, D_2, \ldots\}$ where all these random variables are mutually independent. The subsequence $(U_n, n = 1, 2, \ldots\}$ is called "Up Times," and the subsequence $\{D_n, n = 1, 2, \ldots\}$ is called "Down Times." Each one of these subsequences consist of positive i.i.d. random variables, and hence these are renewal sequences. The combined alternating sequence is called *alternating renewal sequence*. The U variables have a common distribution F_U and the D variables have a common distribution G_D. We are interested in the distribution of the total Up-Time in an interval $(0, t)$. We define this random variable as

$$W(t) = \int_0^t I\{s \epsilon U\}ds. \qquad (11.26)$$

Notice that $0 < W(t) \leq t$. $W(t) = t$ if and only if $U_1 > t$. The fraction of time that the process is in a U-state is $A(t) = \frac{W(t)}{t}$. People often apply the asymptotic result $\lim_{t \to \infty} A(t) = \frac{E\{U\}}{E\{U\} + E\{D\}}$. Later, we will see how close is the asymptotic result to exact evaluation.

Let $N_U(w)$ be the number of U periods within the interval $(0, w)$. Define the *compound renewal process*

$$Y^*(w) = \sum_{n=0}^{N_U(w)} D_n.$$ (11.27)

The random variable $W(t)$ is the stopping time

$$W(t) = \inf\{w > 0 : Y^*(w) \geq t - w\}.$$ (11.28)

Notice that for $0 < w < t$, $P\{w < W(t) < t\} = H^*(t - w, w)$, where

$$H^*(t - w, w) = \sum_{n=0}^{\infty} \left[F_U^{(n)}(w) - F_U^{(n+1)}(w) \right] G_D^{(n)}(t - w).$$ (11.29)

The density of $W(t)$ in the interval $(0, t)$ has two components. One corresponds to the case where $Y^*(w) = (t - w)$, and the second one to the case where $Y^*(w) > t - w$. The first case occurs when the process at time w is in an Up-Time. The second case is when the process at time w is in a Down-Time. Let $\psi_W(w; t) = \psi_W^{(1)}(w; t) + \psi_W^{(2)}(w; t)$ denote the density of $W(t)$ at w. We have

$$\psi_W^{(1)}(w; t) = \sum_{n=1}^{\infty} \left[F_U^{(n)}(w) - F_U^{(n+1)}(w) \right] g_D^{(n)}(t - w),$$ (11.30)

and

$$\psi_W^{(2)}(w; t) = \sum_{n=0}^{\infty} \left[f_U^{(n+1)}(w) - f_U^{(n)}(w) \right] G_D^{(n)}(t - w).$$ (11.31)

The moments of $W(t)$ are

$$E\{W^k(t)\} = t^k(1 - F_U(t)) + k \int_0^t w^{k-1} H^*(t - w, w) dw.$$ (11.32)

In the special case that F_U is exponential with mean $\frac{1}{\lambda}$ and G_D is exponential with mean $\frac{1}{\mu}$ we obtain

$$H^*(t - w, w) = \sum_{n=0}^{\infty} p(n; \lambda w)(1 - P(n - 1; \mu(t - w)))$$

$$= \sum_{j=0}^{\infty} p(j; \mu(t - w)) P(j, \lambda w).$$ (11.33)

In Table 11.6 we present the values of $\psi_W(w; t)$ for the exponential case with $\lambda = 1$, $\mu = 2$, $t = 10$.

Table 11.6 Values of the
density of $\psi_W(w; 10)$

w	$\psi_W(w; 10)$
1	0.00000544
2	0.00025082
3	0.00372696
4	0.02678825
5	0.10729610
6	0.25005400
7	0.33119780
8	0.22352930
9	0.05608870
10	$\exp\{-10\}$

The expected value and the variance of $W(t)$ are: $E\{W(10)\} = 6.77778$ and $V\{W(10)\} = 1.39506$. Thus, $A(10) = 0.68$ while the asymptotic result is $1/(1 + 0.5) = 0.67$.

11.4 Compound Poisson Process Subordinated by a Poisson Process

There are situations in which a compound Poisson process $Y(t)$ cannot be observed continuously in time, but only at random times, according to an independent Poisson process $N(t)$. In such a case we say that $Y(t)$ is subordinated by $N(t)$. Such situations might be prevalent in financial markets, in inventory control, or in industrial applications. Formally, consider the compound Poisson process $Y(t) = \sum_{n=0}^{M(t)} X_n$, where $M(t)$ is a Poisson process with intensity μ. Let $N(t)$ be a Poisson process with intensity λ, independent of $Y(t)$. The process $Z(t) = Y(N(t))$ is called a compound Poisson process subordinated by $N(t)$. Notice that

$$Z(t) = \sum_{n=0}^{N(t)} Y_n, \tag{11.34}$$

where $\{Y_n, n \geq 1\}$ is a sequence of i.i.d. random variables, independent of $N(t)$, distributed like $Y(1)$. Thus $Z(t)$ is also a compound Poisson process. Let $H_Y(y, t)$ be the c.d.f. of $Y(t)$ and $H_Z(z, t)$ that of $Z(t)$. Accordingly,

$$H_Z(z, t) = \sum_{n=0}^{\infty} p(n, \mu t) H_Y(z, n)$$

$$= \exp(-\mu t(1 - e^{-\lambda})) \left[I\{z \geq 0\} e^{-\lambda z} + \sum_{n=1}^{\infty} \left(\frac{\lambda^n}{n!} \right) B_n(\mu t e^{-\lambda t}) F_X^{(n)}(z) \right].$$

$$\tag{11.35}$$

Table 11.7 Bell
polynomials, $n = 1, \ldots, 5$

n	B_n
1	x
2	$x + x^2$
3	$x + 3x^2 + x^3$
4	$x + 7x^2 + 6x^3 + x^4$
5	$x + 15x^2 + 25x^3 + 10x^4 + x^5$

The function $B_n(x)$ is called a Bell polynomial of order n, which is equivalent to the n-th moment of a Poisson with mean x. It can be computed recursively by

$$B_0(x) = 1$$
$$B_{n+1}(x) = x \left[B_n'(x) + B_n(x) \right], n \geq 0. \tag{11.36}$$

For example, in Table 11.7 we show some of these polynomials
The p.d.f. of $Z(t)$ can be written in the form

$$h_Z(z, t) = \sum_{n=1}^{\infty} p_n(t) f_X^{(n)}(z), \tag{11.37}$$

where

$$p_n(t) = \exp\left(-\mu t(1 - e^{-\lambda})\right) \left(\frac{\lambda^n}{n!}\right) B_n\left(\mu t e^{-\lambda t}\right). \tag{11.38}$$

The Laplace transform of $Z(t)$ is

$$E\{\exp\{-\theta Z(t)\}\} = \exp\left\{-\mu t(1 - e^{-\lambda(1 - \psi_X(\theta))})\right\}, \tag{11.39}$$

where $\psi_X(\theta)$ is the moment generating function of X. In particular,

$$E\{Z(t)\} = \lambda \mu \xi t, \tag{11.40}$$

and

$$V\{Z(t)\} = \lambda \mu \left(\sigma^2 + (1 + \lambda)\xi^2\right) t. \tag{11.41}$$

As before, $\xi = E\{X\}$ and $\sigma^2 = V\{X\}$.
For further reading see the paper of Di Crescenzo, Martinucci, and Zacks (2015).

11.5   Telegraph Processes

The material in this section is based on a paper of Zacks (2004). A telegraph process
is a special alternating renewal process, which is applied to describe alternating
movements up and down. We introduced this process in Chap. 4, when we derived
the distribution of the time until particles moving up and down at random are
absorbed in the origin.

Assume that the initial movement is from the origin up. The first movement is
to the positive direction, at velocity $V(t) = c$ and at random time duration U_1.
At time $t = U_1$ the particle changes to the negative direction (down), at velocity
$V(U_1+) = -d$ for a random time D_1. At time $t = U_1 + D_1$ the direction of
movement is changed again, and so forth.

Let $X(t)$ denote the location of the particle on the real line at time t. Let $S_n =
U_n + D_n$, where $S_0 = 0$. Then

$$X(t) = \sum_{n=0}^{\infty} I\{S_n < t < S_n + U_{n+1}\}\, [S_n(c-d) + c(t-S_n)]$$

$$+ I\{S_n + U_{n+1} < t < S_{n+1}\}\, [S_n(c-d) + cU_{n+1} - d(t-(S_n + U_{n+1}))]$$

$$= W(t)c - (t - W(t))d,$$

(11.42)

where $W(t)$ is defined in (11.26). Accordingly,

$$P\{X(t) \le x\} = P\left\{ W(t) \le \frac{x+td}{c+d}; t \right\}.$$

(11.43)

Note that the support of $X(t)$ is within the interval $(-dt, ct]$. The density of $X(t)$
is

$$f_X(x; t) = \left(\frac{1}{c+d}\right) \psi_W\left(\frac{x+td}{c+d}; t\right).$$

(11.44)

The expected value of $X(t)$ is

$$E\{X(t)\} = (c+d)E\{W(t)\} - td.$$

(11.45)

The variance is

$$V\{X(t)\} = (c+d)^2 V\{W(t)\}.$$

(11.46)

Stadje and Zacks (2004) studied the telegraph process when the velocity is
chosen at random after each turning point. Di Crescenzo, Martinucci, and Zacks
(2013) studied a generalized telegraph process with random jumps at each turn
(Table 11.8).

Table 11.8 Density values
of $X(t)$ for exponential
distributions with $\lambda = 1, \mu = 1, t = 10, c = 1, d = 2$

x	$f(x, 10)$	x	$f(x, 10)$
-19	0.000139	-4	0.084026
-18	0.000506	-3	0.081608
-17	0.001327	-2	0.076012
-16	0.002885	-1	0.067848
-15	0.005503	0	0.057937
-14	0.009496	1	0.047219
-13	0.015101	2	0.036609
-12	0.022407	3	0.026880
-11	0.031288	4	0.018573
-10	0.041364	5	0.011971
-9	0.052011	6	0.007104
-8	0.062402	7	0.003802
-7	0.071612	8	0.001772
-6	0.078738	9	0.000669
-5	0.093022	10	0.000166

11.6 Applications

In the present section we describe and analyze several applications of the theory outlined in the previous sections. We start with applications in inventory theory, and then proceed to queuing theory, insurance theory, and conclude with telegraph processes.

11.6.1 Inventory Control

A certain item is stocked for the possible purchase of customers. It is assumed that the customers arrive at the shop independently, following a homogeneous Poisson process with intensity λ. Customers buy random quantities of the item $X_1, X_2, \ldots$, which are positive independent random variables having a common distribution F. The quantities purchased are independent of the arrival times. Thus, the compound Poisson process $Y(t) = \sum_{n=0}^{N(t)} X_n$ presents the total number or quantities of items purchased at the time interval $(0, t]$. We consider in the present example the problem with perishable commodities. More specifically, suppose that the material in stock has to be kept in special condition (cold temperature) and has to be discarded after t_0 time units (for example, bottles of milk). If the arrival intensity λ is high, the expected number of units bought before discarding time will be high, and most of the units will be purchased on time. Items which are not purchased on time (before t_0) are discarded at a loss. If all items are purchased before t_0 replenishment of stock is done immediately, at some replenishment cost. The question is how many units, q, should be stored.

The optimal value of q is the one for which the expected operation cost is minimal. This optimal quantity depends on the cost of replenishment, the cost of holding material in storage, the loss due to discarding, the arrival intensity λ, and the distribution F of the quantities purchased by individuals. We show here some of the required functionals for the optimization.

The time between replenishments of stock is a stopping time $T^*(q) = \min\{T_0(q), t_0\}$, where $T_0(q) = \inf\{t > 0 : Y(t) \geq q\}$. The c.d.f. of $T^*(q)$ is

$$\Psi^*(x; q) = I\{x < t_0\}(1 - H(q, x)) + I\{x = t_0\}H(q, t_0). \tag{11.47}$$

The density of $T^*(q)$ in the interval $(0, t_0)$ is

$$\psi^*(x; q) = \lambda e^{-\lambda x}(1 - F(q)) + \lambda \int_0^q h(y, x)(1 - F(q - y))dy. \tag{11.48}$$

The expected value of $T^*(q)$ is

$$R\{T^*(q)\} = \int_0^{t_0} H(q, t)dt. \tag{11.49}$$

The expected discarded quantity is $E\{D(q)\} = qH(q, t_0)$. Suppose that the holding cost of items in store is \$c per unit item per unit time. Then the total holding cost between replenishments is

$$E\{K(q)\} = cq\,E\{T^*(q)\} - c \int_0^{t_0*} \psi^*(t; q) \int_0^t E\{Y(s)|T^*(q) = t\}dsdt. \tag{11.50}$$

Notice that

$$E\{Y(s)|T^*(q) = t\} = \frac{\int_0^q yh(y, s)h(q - y, t - s)dy}{\int_0^q h(y, s)h(q - y, t - s)dy}. \tag{11.51}$$

In Table 11.9 we illustrate $E\{Y(s)|T^*(q) = t\}$ for $t = 10$, when F and G are exponentials with parameters $\lambda = \mu = 1$ and $q = 10$.

The conditional expected holding cost is $50c$.

11.6.2 Queuing Systems

One of the important characteristics of a queuing system is the busy period of a one-server service station. Customers arrive at a service station one by one at random times. Each customer requires a random amount of service. Customers who arrive while other one is served wait in queue. The length of time the server is continuously

Table 11.9
$E\{Y(s)|T^*(q) = t\}$, with
$q = 10, t = 10$

s	$E\{Y(s)|T^*(10) = 10\}$
1	1.5352
2	2.2578
3	3.1074
4	4.0359
5	5.0000
6	5.9641
7	6.8926
8	7.7422
9	8.4647
10	10.0000

busy with customers is called the *busy period*. We develop here the distribution of the length of the busy period, for the case that arrival of customers follow a Poisson process $N(t)$ with intensity λ, and their required service times $X_1, X_2, \ldots$ are i.i.d. random variables having a common distribution F. The compound Poisson process $Y(t) = \sum_{n=0}^{N(t)} X_n$ designates the total amount of service required in the time interval $(0, t)$.

Let X_1 be the service requirement of the first customer in a busy period. If no customer arrives before the service of the first customer is finished, then the length of the busy period is X_1. On the other hand, if customers arrive before the service of the first one is finished, then the conditional density of the length of the busy period is according to (11.26)

$$f_{BP}(t|X_1) = \left(\frac{X_1}{t}\right) h(t - X_1, t), t > X_1. \tag{11.52}$$

Thus, the distribution (c.d.f.) of the busy period is

$$F_{BP}(t) = \int_0^t f_X(x) \left[e^{-\lambda x} + \int_x^t \left(\frac{x}{s}\right) h(s - x, t) ds \right] dx. \tag{11.53}$$

In Table 11.10 we present the conditional c.d.f. of the busy period, given X_1 for the exponential distribution F with parameter μ.

In Table 11.11 we present the total distribution of the busy period, for the exponential F with $\lambda = 1, \mu = 2$.

11.6.3 Application in Insurance

The increasing linear boundary with positive intercept has an application in insurance in determining the *Ruin Probability* of a company. Suppose that an insurance

Table 11.10 Conditional
distribution of the busy
period, given
$X_1, \lambda = 1, \mu = 2$

t	$X_1 = 3$	$X_1 = 5$	$X_1 = 7$
3	0.04979		
4	0.32232		
5	0.51873	0.00674	
6	0.65371	0.11983	
7	0.74764	0.26721	0.00091
8	0.81320	0.40682	0.03980
9	0.86042	0.52619	0.12132
10	0.89472	0.62389	0.22432
11	0.91996	0.70222	0.33192
12	0.93872	0.76439	0.43422
13	0.95280	0.81349	0.52369
14	0.96345	0.85221	0.60675
15	0.97156	0.88272	0.67533
20	0.99143	0.96219	0.88051
25	0.99924	0.98734	0.95686

Table 11.11 Distribution of
the busy period

t	$F_{BP}(t)$
0.1	0.17332
0.2	0.30424
0.3	0.40508
0.4	0.48424
0.5	0.54748
0.6	0.59886
0.7	0.64124
0.8	0.67668
0.9	0.70667
1.0	0.73241
1.5	0.81970
2.0	0.86968
2.5	0.90166
3.0	0.92361
3.5	0.93940
4.0	0.95615
4.5	0.96012
5.0	0.96711

company starts with a capital C, and the income from premiums is 1 per unit time.
Let the compound Poisson process $Y(t)$ designate the total payments for claims in
the time interval $(0, t)$. The first time $Y(t)$ crosses the boundary $B(t) = C + t$ is
called a ruin time. This is the stopping time

$$T_C = \inf\{t > 0 : Y(t) > C + t\}. \tag{11.54}$$

Table 11.12 Ruin
probabilities in the
exponential case

λ	μ	C	RP
1	2	0.05	0.4756
	2	0.5	0.3033
	2	1	0.1839
	2	2	0.0678
1	1.2	0.05	0.6463
5	6	0.5	0.3483
		1	0.2042
		2	0.0595

The probability that $\{T_C < \infty\}$ is called the ruin probability. Let X denote the amount paid for a claim, and let $\rho = \lambda E\{X\}$. If $\rho > 1$, then $P\{T_C < \infty\} = 1$, which means that ruin will occur sooner or later for sure. As shown by Perry, Stadje, and Zacks (2002, p. 59) when $\rho < 1$, the ruin probability is

$$P\{T_C < \infty\} = (1 - \rho) \int_0^\infty h(t + C, t)dt. \qquad (11.55)$$

The following results are obtained from this formula (11.55) for the case of $X \sim \exp(\mu)$.

By using Laplace transforms we obtain for the exponential case an exact formula (see Asmussen and Albrecher 2010)

$$P_E\{T_C < \infty\} = \rho \exp\{-\mu(1 - \rho)C\}. \qquad (11.56)$$

With this formula we get for $\lambda = 1, \mu = 2, C = 0.05, P_E = 0.4756$. This is equal to the value in the Table 11.12.

Chapter 12
Current Challenges in Statistics

Much has been written on the future of statistics in the era of rapid technological innovations, and huge data quantities that require quick analysis. The number of statistics and biostatistics departments at universities is growing. The number of Ph.D. degrees conferred to students is in the hundreds, and there are many positions available to statisticians in industries, hospitals, business firms, and universities. Although it seems that the state of Statistical Science is strong, it is challenged now by the trend to develop departments of Data Science. In the past, departments of Computer Science focused attention on teaching hardware engineering or software engineering. Their students were directed at many universities to study statistics and applied probability in statistics or mathematics departments. In the future, this cooperation might end. Even among statisticians there are calls to change the way we teach statistics from models to ready-made algorithms (see Kaplan 2018). To be a good data scientist one should be strong in mathematics, statistics, applied probability, operations research (optimization), and computers. It seems to me that a BS degree in mathematics and MS degree in applied statistics, with a strong emphasize on big data analysis, like courses on data mining, machine learning, and neuro-networks, might give the students an excellent preparation for data analysis. Algorithms are computational instructions and as such cannot replace models. Science is based on theory and models, if data refutes a theory a new theory is required, or a modification of the old one. Experiments in science laboratories are of moderate size and are based on experimental design. Usually experiments are done in order to test a hypothesis. For this we still need statistical methodology based on models. Thus basic statistical education is required for anyone who analyzes data. In the following section some ideas based on experience are presented about teaching statistics to data scientists. Departments of statistics could serve well departments of data science. This requires creating courses with appropriate syllabuses. This is a real challenge that should be taken seriously.

© Springer Nature Switzerland AG 2020
S. Zacks, *The Career of a Research Statistician*, Statistics for Industry,
Technology, and Engineering, https://doi.org/10.1007/978-3-030-39434-9_12

12.1   Teaching Statistics to Data Scientists

On the basis of notes written for the workshops on quality and reliability given to industry engineers, as described in Sect. 1.12, R.S. Kenett and S. Zacks published in 1998 with Duxbury Press, a book entitled *Modern Industrial Statistics: Design and Control of Quality and Reliability*. The book contained 14 chapters in 2 parts and 70 data sets. MINITAB routines and graphs accompanied the text. S-Plus functions, for each chapter, were given in an appendix, as well as executable (exe) files of Quick Basic programs, which were developed for fast computations of advanced method, like bootstrapping and simulations of advanced techniques. A brief contents of the book is given in order to illustrate what we think is important to teach students of data science.

Topics contained in the first eight chapters of the book, appropriate for data analysis, are: understanding randomness and variability; variability in several dimensions; basic models of probability and distribution functions; sampling for estimation of finite population quantities; distribution free-inference including bootstrapping: role of simulation and computer-intensive techniques; multiple linear regression and analysis of variance.

We see that the above listed topics contain material for a whole semester of intensive study. We have used at the Engineering School of Binghamton University an abridged version of the book, entitled *Modern Statistics: A Computer-Based Approach*, Thompson Learning (2001), for a one semester course. A one semester course on advanced methods of industrial statistics was given in the graduate program of the Mathematical Sciences department at Binghamton University. The course was given based on the second part of the book. The advanced methods included the following topics: nonparametric tests for randomness, based on run tests; modified Shewhart control charts; the economic designs of control charts; multivariate control charts; cumulative sum control charts (CUSUM); average run length, probability of false alarm and conditional expected delay; Bayesian detection of change-points; Shiryaev–Roberts procedures; process tracking. The last topics were discussed at length in Chap. 2 of the present monograph. The *Modern Industrial Statistics: Design and Control of Quality and Reliability* textbook was translated to Spanish, Thompson International (2000). A Chinese edition was published by China Statistics Press (2003), with cover pages, preface, and table of contents translated to Chinese, while the rest of the book was left in English. A soft cover edition was published by Brooks-Cole (2004). The book was used for 15 years at various universities.

A second edition of the book was published in 2013 in the Wiley series *Statistics in Practice* under the slightly different title *Modern Industrial Statistics With Applications in R, MINITAB and JMP*. The second edition contained 15 chapters in 5 parts. It contains a new chapter on Computer Experiments. The software S-PLUS was replaced by R, and the software JMP was added for analyzing examples. Special R packages were developed for the book and are available online for the use of the readers from the dedicated website www.wiley.com/go/modern_industrial_statistics. This second edition was translated to Vietnamese.

Another book resulting from workshops on reliability methods is that of Zacks (1992), entitled *Introduction to Reliability Analysis, Probability Models and Statistical Methods,* Springer-Verlag. The emphasis in these books is on computerized teaching of statistics, and they could serve well short courses in statistics for data science students.

12.2 Additional Comments About Data Analysis

According to the title of a new book by Kenett and Redman (2019), the real work of data science is to turn data into information with which people can do better decisions, and organizations become stronger. The analysis should be rooted in sound theory, otherwise solutions will not be relevant for long. Real data is partly deterministic and mostly random. In order to predict outcomes with high degree of confidence and precision one has to explore the behavior of the random components. Probability models and statistical goodness of fit methods can serve well in this regard. See for example the MCCRES data analysis in Chap. 5. In addition, sequential sampling can be useful in many cases, as well as detection of change-points or outliers should be considered.

A big data set can be considered as a big sample from some "super population." Thus, data are realization of random variables. Functions of random variables are estimators if they estimate deterministic components (nonrandom parameters). One cannot estimate random variables. If the quantities of interest are random, we try to predict their future realization based on their past realizations. These should be chosen so as to minimize the corresponding prediction risk. The basic theory for prediction in finite populations was given in Chap. 6. In Chap. 5 of the book entitled "Stage-Wise Adaptive Designs" (Zacks 2009) we can read about one day ahead forecasting of time-series. The type of predictors discussed there are: Linear predictors for covariance-stationary time-series; Quadratic LSE predictors for nonstationary time-series; Moving average predictors for nonstationary time-series; Predictors for general trends with exponential smoothing; Dynamic linear models predictors for nonstationary time-series. All these predictors are based on the mathematical theory of time-series. Their performance is illustrated graphically on interesting examples. The R-programs of all these predictors are given in the Appendix of the book. The tracking procedures of Chap. 2 of the present book can also be applied for forecasting the output of a system. These predictors are model driven.

Repeatability of analytical results from prediction models should be established by applying the regression predictor from past data on new data sets. Repeatability may not be guaranteed by the usual cross-validation techniques applied on the same data set.

To summarize, the challenges to the statistical communities in the era of huge data sets are many and require much thinking, theoretical modeling, and flexible approach.

Appendix: R Programs

Most of the tables and all the figures in this monograph were calculated according to the R programs given here. These R functions are presented according to the chapter in which they are used. Functions which are in the library of R are used without comments. The reader is advised to insert the following functions to his/her collection of objects

```
1. std(x){
sqrt(var(x))}
2. inv(x){
solve(x)}
3. jay(k,m){
l= k*m
a= c(1:l)/c(1:l)
res= matrix(a,nrow=k)
res}
4. psum(x){
n=length(x)
res=c(1:n)
res[1]=x[1]cc
for(i in 2:n){
res[i]=res[i-1]+x[i]
}
res
}
```

The function std(x) computes the standard deviation of a data set x.
The function Inv(x) computes the inverse of a non-singular matrix x.
The function jay(k,m) creates a kxm matrix of 1's.
The function psum(x) computes the partial sums of the data set x.

© Springer Nature Switzerland AG 2020
S. Zacks, *The Career of a Research Statistician*, Statistics for Industry,
Technology, and Engineering, https://doi.org/10.1007/978-3-030-39434-9

A.1   Chapter 2

Figure 2.1 was created with the following R Function "track." The input variables
of this function are the parameters:

mu0=μ_0; taus=τ^2; sigse=σ_ϵ^2; p=p; zeta=ς; sigsz=σ_z^2; M=# of tracking points.

The function "track" given below depends on the lower triangular matrix given
by the function

```
trian←
function(n){
res<-jay(n,n)
if(n>1){
for(i in 1:(n-1)){
for(j in (i+1):n){
res[i,j]<-0}}}
res
}

track=
function(mu0,taus,sigse,p,zeta,sigsz,M){
resmu=c(1:M)
resJ=rbinom(M,1,p)
resZ=rnorm(M,zeta,sqrt(sigsz))
mu1=rnorm(1,mu0,sqrt(taus))
resX=c(1:M)
resmuh=c(1:M)
resmu[1]=mu1
resmu[2:M]=mu1*jay(1,M-1)+psum(resJ[1:(M-1)]*resZ[1:(M-1)])
resX=resmu+rnorm(M,0,sqrt(sigse))
plot(c(1:M),resX,xlab="n",ylab="X_n")
resmuh[1]=resX[1]
v1=taus+sigse
v2=p*sigsz+p*(1-p)*(zeta^2)
ubound=c(1:M)
lbound=c(1:M)
for(n in 2:M){
Vn=0*jay(n,n)
Vn[1,1]=v1
Vn2=v1*eye(n-1)+v2*(trian(n-1)%*%t(trian(n-1)))
Vn[2:n,2:n]=Vn2
Xn=c(1:n)
for(i in 1:n){
Xn[i]=resX[n-i+1]}
resmuh[n]=(t(Xn)%*%inv(Vn)%*%jay(n,1))/(jay(1,n)%*%inv(Vn)%*%jay(n,1))
ubound[n]=mu0+4/sqrt(jay(1,n)%*%inv(Vn)%*%jay(n,1))
```

```
lbound[n]=mu0-4/sqrt(jay(1,n)%*%inv(Vn)%*%jay(n,1))
}
ubound[1]=mu0
lbound[1]=mu0
points(c(1:M),resmuh)
lines(c(1:M),resmuh)
lines(c(1:M),ubound)
lines(c(1:M),lbound)
out=list(resmu,resmuh)
out
}
```

For creating Fig. 2.2 we need the function "amocpois." For this function we need the p.d.f. of Negative-Binomial. This is given by

```
dnegbinom=
function(j,p,k){
res=gamma(j+k)*((1-p)^k)*(p^j)/(gamma(j+1)*gamma(k))
res
}
```

The input variables for the function "amocpois" are: x= a random sequence of Poisson r,v,;nw=length of window; al1=mean of Poisson before change; al2=mean of Poisson after change;v1,v2,p are prior parameters of negative binomial.

```
amocpois=
function(x,nw,al1,al2,v1,v2,p){
Ns=length(x)
plot(c(1:Ns),x,xlab="n",ylab="X")
res=c(1:Ns)
res[1]=x[1]
for(it in 2:Ns){
n=min(it,nw)
itn=it-n+1
xn=x[itn:it]
sn=sum(xn)
pst=0*c(1:n)
for(j in 1:(n-1)){
sj=sum(xn[1:j])
snj=sn-sj
fj=(j*al1)/(v1+j*al1)
gj=dnegbinom(sj,fj,v1)
fnj=((n-j)*al2)/(v2+(n-j)*al2)
gnj=dnegbinom(snj,fnj,v2)
pst[j]=p*((1-p)^(j-1))*gj*gnj
}
fn=n*al1/(v1+n*al1)
```

```
pst[n]=((1-p)^(n-1))*dnegbinom(sn,fn,v1)
poster=pst/sum(pst)
mut=0
for(j in 1:(n-1)){
sj=sum(xn[1:j])
snj=sn-sj
mut=mut+(poster[j]*al2*(snj+v2))/(v2+(n-j)*al2)
}
mut=mut+(poster[n]*al1*(v1*sn))/(v1+n*al1)
res[it]=mut
}
res
lines(c(1:Ns),res)
}
```

The following function simulates the Shiryaev–Roberts detection procedure.

```
SHRO=
function(al0,al1,n0,alf,NS){
res=0*c(1:100)
resn=c(1:NS)
resw=(1:NS)
r0=al1/al0
dlt=al1-al0
cr=(1-alf)/alf
for(i in 1:NS){
n=1
x=rpois(1,al0)
rx=(r0^x)*exp(-dlt)
W=rx
res[1]=W
repeat{
if((W>cr)||(n>10000))
break
n=n+1
if(n<=n0){x<-rpois(1,al0)}
if(n>n0){x<-rpois(1,al1)}
rx=(r0^x)*exp(-dlt)
W=(1+W)*rx
res[n+1]=W
}
resn[i]=n
resw[i]=W}
med=quantile(resn,prob=0.5)
dec=quantile(resn,prob=0.1)
hist(resn)
```

```
out=list(dec,med,max(resn),mean(resn),mean(resw))
out
}
```
The following function simulates the CUSUM procedure in the Poisson case.
```
cuspois=
function(al1,al2,alf,tho,Ns){
K=(al2-al1)/log(al2/al1)
h=-log(alf)/log(al2/al1)
resn=c(1:Ns)
resp=0*c(1:Ns)
resc=0*c(1:Ns)
for(i in 1:Ns){
n=1
if(n<=tho){x<-rpois(1,al1)}
if(n>tho){x<-rpois(1,al2)}
S=max(0,x-K)
repeat{
if(S>h)
break
n=n+1
if(n<=tho){x<-rpois(1,al1)}
if(n>tho){x<-rpois(1,al2)}
S=max(0,S+max(0,x-K))}
resn[i]=n
if(n<=tho){resp[i]<-1}
if(n>tho){resc[i]<-n}}
hist(resn)
ced=sum(resc)/(Ns-sum(resp))-tho
out=list(mean(resn),std(resn),mean(resp),ced)
out
}
>
```

A.2   Chapter 3

Tables 3.1 and 3.2 were computed by numerical integration. Table 3.3 presents simulation estimates computed with the function "simmuB." This function uses the functions "fen" and "fen1" which are given below:

```
fen=
function(u,V,Ws,n){
f1=u^((n-1)/2)
```

```
f2=(1-u)^(n/2)
f3=exp(-(Ws^2)*(1-u)/(2*n))/sqrt((1-u)/n)
f4=(u+V*(1-u))^(n-1)
res=f1*f2*f3/f4
res
}

fen1=
function(u,V,Ws,n){
f1=u^((n-1)/2)
f2=(1-u)^(n/2)
f3=exp(-(Ws^2)*(1-u)/(2*n))/sqrt((1-u)/n)
f4=(u+V*(1-u))^(n-1)
res=u*f1*f2*f3/f4
res
}

simmuB=
function(n,rho,Ns){
resmu=c(1:Ns)
resq=c(1:Ns)
reszb=c(1:Ns)
for(j in 1:Ns){
V=rho*rf(1,n-1,n-1)
XB=mean(rnorm(n,0,1))
YB=mean(rnorm(n,0,sqrt(rho)))
Ws=YB-XB
g1=0
g=0
for(i in 1:8){
g1=g1+fen1(z[i],V,Ws,n)*w[i]
g=g+fen(z[i],V,Ws,n)*w[i]
}
EB=g1/g
muB=EB*XB+(1-EB)*YB
reszb=EB
resmu[j]=muB
resq[j]=muB^2
}
out=list(mean(reszb),mean(resmu),var(resmu))
out
}
```

Table 3.4 was computed according to (3.15), by numerical integration.

Table 3.5 was computed by the function "simken" given below:

```
simken=
function(rho,n,Ns){
res1=c(1:Ns)
res2=c(1:Ns)
for(i in 1:Ns){
XB=rnorm(1,0,sqrt(1/n))
YB=rnorm(1,0,sqrt(rho/n))
Q1=rchisq(1,n-1)
Q2=rho*rchisq(1,n-1)
sd=sqrt((1+rho)/n)
u2=rnorm(1,0,sd)
z1=Q1/u2^2
z2=Q2/u2^2
k=(n-1)/2
s=0
s1=0
for( j in 1:8){
u=z[j]
aj=(1/(1-u)^2)*((1-u)^k/u^k)*sqrt(1-u)*(z1+z2*(1-u)/u+n*(1-u))^(-n+3/2)
aj=aj*exp(-u/(2*(1-u)))/2
s=s+aj*w[j]
aj1=(1/(1-u))*((1-u)^k/u^k)*sqrt(1-u)*(z1+z2*(1-u)/u+n*(1-u))^(-n+3/2)
aj1=aj1*exp(-u/(2*(1-u)))/2
s1=s1+aj1*w[j]
}
FB=s1/s
res1[i]=FB
mu=XB*(1-FB)+YB*FB
res2[i]=mu
}
out=list(mean(res1),var(res1),mean(res2),var(res2))
out
}
```

Table 3.6 was computed with the function "SimBC," which is

```
simBC=
function(rho,n,Ns){
res=c(1:Ns)
for(i in 1:Ns){
XB=rnorm(1,0,sqrt(1/n))
YB=rnorm(1,0,sqrt(rho/n))
R=rho*rchisq(1,n-1)/rchisq(1,n-1)
```

```
D=YB-XB
res[i]=XB+D*1.39/(1+(n-1)*R/(n+2)+(D^2)/(n+2))
}
out=list(mean(res),var(res))
out
}
```

A.3   Chapter 4

Function "surv5" below was used to compute $P\{S^{(5)}|K_4 = 4\}$, with the parameters $\lambda = 2.154613$, $n = 20$, and $p = 0.7$,

```
surv5=
function(n,al,p){
x=1-p
s=0
s1=0
for(j in 4:n){
pr=1-(x^(j-3))*(1+x+x^2+x^3)+(x^(2*j-5))*(1+x+2*x^2+x^3+x^4)
pr=pr-(x^(3*j-6))*(1+x+x^2+x^3)+x^(4*j-6)
s=s+pr*dpois(j,al)
s1=s1+pr*(x^(j-4))*dpois(j,al)
}
out=s1/s
out
}
```

The following is the program to compute Table 4.2(b) with

```
rnegbin=
function(p){
u=rbinom(1,1,p)
n=1
repeat{
if(u==1)break
n=n+1
u=rbinom(1,1,p)}
out=n
out
}

ElasticC2=
function(al,mu,alf,Ns){
resC=c(1:Ns)
resA=c(1:Ns)
```

```
for(i in 1:Ns){
U=rgamma(1,1)/al
D=rgamma(1,1)/mu
repeat{
if(D>U) break
U=U+rgamma(1,1)/al
D=D+rgamma(1,1)/mu}
C=2*U
resC[i]=C}
for(i in 1:Ns){
m=rnegbin(alf)
s=sample(resC,m,replace=TRUE)
resA[i]=sum(s)}
out=ist(mean(resC),var(resC),mean(resA),var(resA))
out
}
```

A.4   Chapter 5

The function "Bayris" computes the optimal value k^0 and the minimal risk $\varrho^0(\varphi, \upsilon)$

```
Bayrisk=
function(p,c,bet,v){
gam=p/(c+p)
fee=bet/(1+bet)
k0=qnegbinom(gam,fee,v)+1
s0=0
for(j in 0:(k0-1)){
s0=s0+j*dnegbinom(j,fee,v)}
r0=p*v*fee/(1-fee)-(c+p)*s0
out=list(k0,r0)
out
}
```

Table 5.1 was computed with the function "Tbayriskex" which uses the function "Mbayriskex." These functions are:

```
Mbayriskex=
function(p,c,bet,v,n,Ns){
fee=bet/(1+n*bet)
s=0
for(j in 1:n){
s=s+rnbinom(1,v,1-fee)}
st=0
for(i in 1:Ns){
```

```
st=st+Bayrisk(p,c,bet,v+s)*dnbinom(i-1,v+s,1-fee)}
out=st
out}

Tbayriskex=
function(p,c,bet,v,n,Ns){
tres=c(1:n)
for(j in 1:n){
betj=bet/(1+j*bet)
MB=Mbayriskex(p,c,betj,v,j,Ns)
tres[j]=MB}
out=Bayrisk(p,c,bet,v)+sum(tres)
out}
```

A.5   Chapter 8

The function flm computes the probability of hitting the lower boundary in Truncated Case 1, and the OC function

```
flm=
function(k1,k2,al){
tm=c(k1:(k2+k1))
pm=c(0:k2)
pm[1]=exp(-al*k1)
for(m in 1:k2){
pm[m+1]=dpois(m,al*tm[m+1])
sm=0
for(j in 1:m){
sm=sm+pm[j]*dpois(m+1-j,al*(m+1-j))}
pm[m+1]=pm[m+1]-sm
}
pl=sum(pm)
out=list(pm,pl)
out
}
```

The program surv computes the survival function of $T_s^{(1)}$ given λ, and $E_\lambda\{T_s^{(1)}\}$.

```
surv=
function(k1,k2,al){
tm=0*c(k1:(k2+k1))
pm=0*c(0:k2)
sur=0*c(1:(k1+k2))
pm[1]=exp(-al*k1)
```

```
for(m in 1:k2){
pm[m+1]=dpois(m,al*tm[m+1])
sm=0
for(j in 1:m){
sm=sm+pm[j]*dpois(m+1-j,al*(m+1-j))}
pm[m+1]=pm[m+1]-sm
}
for(i in 1:k1){
sur[i]=ppois(k2-1,al*i)}
for(lt in (k1+1):(k1+k2-1)){
sur[lt]=ppois(k2-1,al*lt)-ppois(lt-k1,al*lt)
sr=0
l=lt-k1
for(j in 0:(l-1)){
sr=sr+pm[k1+j-1]*(ppois(k2-1-j,al*(l-j))-ppois(l-j,al*(l-j)))
}
sur[lt]=sur[lt]-sr
}
sur2=sur[1:(k1+k2-2)]
tx=c(1:(k1+k2))
plot(tx,sur)
out=list(sur,sum(sur2))
out
}
```

The following function computes the defective probability function, $g(j,l)$, for Trunc. 1, g1L

```
g1L=
function(j,l,k1,k2,al){
fr=dpois(j,al*l)
if(l<=k1){out=fr}
if(l>k1){
lk=l-k1
if(j<=lk){out=0}
if(j>lk){
sm=0
for(i in 1:lk){
sm=sm+fL(i,k1,k2,al)*dpois(j-i,al*(lk-i))}
out=fr-sm
}}
out
}
```

The following function, ET1, computes the expected value of $T_s^{(1)}$.

```
ET1=
function(k1,k2,al){
s1=<-0
for(i in 0:(k2-1)){
s1=s1+1-ppois(i,al*k1)}
et=s1/al
s2=0
for(l in 0:(k2-1)){
s3=0
for( j in (l+1):(k2-1)){
s4=0
for(i in 0:(k2-1-j)){
s4=s4+1-ppois(i,al)}
s3=s3+s4*g1L(j,(k1+l),k1,k2,al)}
s2=s2+s3
}
et=et+s2/al
et
}
```

The functions g0L and g20L below are required for computing $\pi(\lambda)$ with pi2

```
g0L=
function(j,l,al){
k=l-j
if(j<l){
s=dpois(0,al*l)
out=s
if(j>0){
for(i in 1:j){
s=s+dpois(i,al*(i+k))*dpois(j-i,al*(l-i))}
out=s
}
}
if(j>=l){out=0}
out
}
g20L=
function(j,l,k1,k2,al){
re=dpois(j,al*l)
if(j<=k2){out<-re}
if(j>k2){
s=0
for(i in 1:(j-k2)){
```

```
s=s+dpois(k2+i,al*i)*g0L(j-k2-i,l-i,al)}
re=re-s
out=re
}
if(j>=(k2+l)){out<-0}
out
}

pi2=
function(k1,k2,ls,al){
ks=k1+k2
ns=(ls-1)*ks
g20=0*c(1:ks)
g21=0*c(1:ks)
g22=0*c(1:ks)
si=0
for(i in 1:(ks-1)){
gi=g20L(i,k1,k1,k2,al)
sx=0
for(j in 1:ks){
sx=sx+gi*flmx(i,ks-i,al)[j]
}
si=si+sx}
re=si+dpois(0,k1*al)
for(i in 1:(ks-1)){
g20[i]=g20L(i,k1,k1,k2,al)
g21[i]=g20[i]
}
for(l in 1:ks){
g22[1]=g21[1]*dpois(0,al)
for(j in 2:ks-1){
sl=0
for(m in 2:j){
sl=sl+g21[m]*dpois(j-m,al)}
g22[j]=sl
}
sl=0
for(m in 2:(ks-1)){
g22[ks]=g21[m]*dpois(ks-m,al)}
g22[ks]=g22[ks]+g21[ks]*dpois(1,al)
g21=g22
}
re}
```

The following functions are for Table 8.9

```
MLEZ=
function(al,mu,bet,zet,n,Ns,Ng){
sm=0*c(1:Ng)
t=c(1:Ng)
e1=0*c(1:Ng)
e2=0*c(1:Ng)
for(i in 1:n){
t[i]=TB(al,mu,bet)
for(l in 1:Ng){
for(j in 0:Ns){
e1[l]=e1[l]+dpois(j,zet-(l-1)*0.05)*dpois(j+1,al*t[i])
e2[l]=e2[l]+dpois(j,zet-(l-1)*0.05)*dpois(j,al*t[i])
}
sm[l]=sm[l]+e1[l]/e2[l]
}
}
M=1
repeat{
if(((sm[M]/n)>1)||(M==Ng))
break
M=M+1
}
out=zet-(M-1)*0.05
out
}

simMLE=
function(al,mu,bet,zet,n,Ns,Ng,NS){
resMLE=c(1:NS)
for(i in 1:NS){
resMLE[i]=MLEZ(al,mu,bet,zet,n,Ns,Ng)}
ma=mean(resMLE)
ss=std(resMLE)
out=list(ma,ss)
hist(resMLE)
out
}
```

A.6 Chapter 10

Program bayesim for computing Table 10.2

```
bayesim←
function(g,p1,xs,xt,n,M,al){
bet0=-3.569
bet1=0.549
res=c(1:(n+1))
yr=c(1:n)
res[1]=xs
w1=p1/(1-p1)
for(i in 1:n){
Tn=0
Td=0
Tb=0
Tq=0
xi=res[i]
qi=exp(bet0+bet1*xi)
qi=qi/(1+qi)
yi=rbinom(1,1,qi)
yr[i]=yi
for(l in 1:M){
b1=rgamma(1,al)
Lj=0
for(j in 1:i){
xj=res[j]
yj=yr[j]
Lj=Lj+log(1+w1*exp(b1*(xj-xs)))-b1*yj*(xj-xs)
}
Ls=exp(-Lj)
Tn=Tn+Ls/b1
Td=Td+Ls
Tb=Tb+Ls*b1
Tq=Tq+Ls*(b1^2)
}
Eb=Tb/Td
Sb=sqrt(Tq/Td-Eb^2)
xgh=-xs+(log(g/(1-g))-log(w1))/(Eb+al*Sb)
xgh=min(xt,xgh)
res[i+1]=xgh
}
res
}
>
```

A.7   Chapter 11

The c.d.f. of the compound Poisson distribution $H(y, t)$ is computed with the function Hn(al,mu,y,t,Ns). The variables in this function are: al=λ, mu=μ, Ns=3 of terms in the sum. In most of the tables we used Ns=100.

```
> Hn=
function(al,mu,y,t,Ns){
s1=0
for(n in 0:Ns){
s1=s1+dpois(n,mu*y)*ppois(n,al*t)}
out=s1
out}
```

The corresponding p.d.f. $h(y, t)$ is computed with

```
> h=
function(al,mu,y,t,Ns){
s=0
for(n in 1:Ns){
s=s+dpois(n,al*t)*mu*dpois(n-1,y*mu)}
out=s
out
}
```

Tables 11.1 and 11.2 were computed with these two functions.

The c.d.f. in Table 11.3 was computed by numerical integration of $\int_5^t (5/t)h(t - 5, t)dt$.

Table 11.4 was computed numerically according to $P\{T_U > t\} = \exp\{-\lambda t\} + \int_0^t \frac{t-y}{t} h(y, t)dy$, with $\lambda = 2, \mu = 1, \beta = 0$.

In Table 11.5 we presented the probabilities $P\{T_U(3) > t\}$. For these computations we used the R function gbt for $g_\beta(y, t)$. This function is given by

```
> gbt=
function(al,mu,y,t,bet,Ns){
res=h(al,mu,y+bet,t,Ns)-h(al,mu,y+bet,y,Ns)*exp(-al*(t-y))
s=0
for(i in 1:8){
x[i]=y*z[i]
s=s+h(al,mu,x[i],x[i],Ns)*((t-y)/(t-x[i]))*h(al,mu,y*(1-z[i]),t-x[i],Ns)*w[i]}
Ia=s*y
res=res-Ia
out=r
es
out
}
```

In Table 11.6 we present the density of $W(t)$, fw(w,t), for $\lambda = 1$, $\mu = 2$, and $t = 10$. For this we used

```
> fw=
function(al,mu,w,t,Ns){
fw1=h(al,mu,t-w,w,Ns)
s=0
for(n in 1:Ns){
s=s+(dpois(n,al*w)-dpois(n-1,al*w))*ppois(n-1,mu*(t-w))
}
fw2=-s*al
res=fw1+fw2
res
}
```

In Table 11.7 we present the density of the location of the telegraph process at time $t = 10$, with velocities $c = 1$. $d = 2$. This density is obtained from fw, i.e., $f_{X(t)}(x) = (1/(c+d)) * f_{W(t)}((x+t*d)/(c+d))$.

The conditional c.d.f. of the Busy Period (BP) given that $\{X_1 = x\}$ is $exp\{-\lambda x\} + x \int_x^t (1/s)H(s-x,s)ds$. These conditional probabilities are computed numerically and are given in Table 11.8. The marginal c.d.f. is given in Table 11.9, and computed according to the R program FBP given below, with $\lambda = 1$, $\mu = 2$.

```
> FBP=
function(al,mu,t,Ns){
R=t
s1=0
for(i in 1:8){
x[i]=R*z[i]
R2=t-x[i]
s2=0
for(j in 1:8){
y[j]=x[i]+R2*z[j]
s2=s2+(x[i]/y[j])*h(al,mu,y[j]-x[i],y[j],Ns)*w[j]
}
In1=R2*s2
s1=s1+mu*exp(-mu*x[i])*(exp(-al*x[i])+In1)*w[i]
}
res=s1*R
res
}
```

The values in Table 11.10 were computed numerically according to the equation $RP = 1 - (1-\rho) \int_0^\infty h(t+C,t)dt$.

References

Anbar, D. (1984). Stochastic approximation methods and their use in bioassay and phase I clinical trials. *Communications in Statistics - Theory and Methods, 13*, 2451–2467.

Asmussen, S., & Albrecher, H. (2010). *Ruin probability* (2nd ed.). Singapore: World Scientific.

Barzily, Z., Marlow, W. H., & Zacks, S. (1979). Survey of approaches to readiness. *Naval Research Logistics Quarterly, 26*, 21–31.

Basu, D. (1969). Role of sufficiency and likelihood principles in survey theory. *Sankhya, Series A, 31*, 441–453.

Basu, D. (1971). An essay on the logical foundations of survey sampling. In V. P. Godambe & D. S. Sprott, (Eds.), *Foundations of statistical inference*. Toronto: Holt, Rinehart and Winston.

Basu, D. (1978). On the relevance of randomization in data analysis. In N. K. Namboodiri (Ed.), *Survey sampling and measurement* (pp. 267–339). New York: Academic.

Bather, A. J. (1963). Control charts and minimization of costs. *Journal of the Royal Statistical Society, B, 25*, 49–80.

Bather, J. A. (1967). On a quickest detection problem. *Annals of Mathematical Statistics, 38*, 711–724.

Bogdanoff, J. L., & Kozim, F. (1985). *Probabilistic models for cumulative damage*. New York: John Wiley.

Bolfarine, H., & Zacks, S. (1991a). Equivariant prediction of the population variance under location-scale superpopulation models. *Sankhya, Series B, 53*, 288–296.

Bolfarine, H., & Zacks, S. (1991b). Bayes and minimax prediction in finite population. *Journal of Statistical Planning and Inference, 28*, 139–151.

Bolfarine, H., & Zacks, S. (1992). *Prediction theory for finite populations*. New York: Springer.

Bolfarine, H., & Zacks, S. (1994). Optimal prediction of the finite population regression coefficient. *Sankhya, Series B, 56*, 1–10.

Bolfarine, H., Rodriguez, J., & Zacks, S. (1993). Some asymptotic results in finite populations. *Statistics, 24*, 359–370.

Breir, S. S., Zacks, S., & Marlow, W. H. (1986). An application of empirical Bayes techniques to the simultaneous estimation of many probabilities. *Naval Research Logistics Quarterly, 33*, 77–90.

Breiman, L. (2001). Statistical modeling: The two cultures. *Statistical Science, 16*, 199–231.

Brown, L. D., & Cohen, A. (1974). Point and confidence estimation of a common mean and recovery of inter-block information. *The Annals of Statistics, 2*, 963–976.

Brown, M., & Zacks, S. (2006). A note on optimal stopping for possible change in the intensity of an ordinary Poisson process. *Statistics & Probability Letters, 76*, 1417–1425.

© Springer Nature Switzerland AG 2020

S. Zacks, *The Career of a Research Statistician*, Statistics for Industry, Technology, and Engineering, https://doi.org/10.1007/978-3-030-39434-9

Chernoff, H., & Zacks, S. (1964). Estimating the current mean of a normal distribution which is subject to changes in time. *Annals of Mathematical Statistics, 35*, 999–1018.

Cochran, W. G. (1977). *Sampling techniques* (3rd ed.). New York: Wiley.

Cohen, A. (1976). Combining estimates of location. *Journal of the American Statistical Association, 71*, 172–175.

Cohen, A., & Sackrowitz, H. B. (1974). On estimating the common mean of two normal distributions. *The Annals of Statistics, 2*, 1274–1282.

Datta, S., & Mukhopadhyay, N. (1995). On fine-tune bounded risk sequential point estimation of the mean of exponential distribution. *South African Statistical Journal, 29*, 9–27.

De, S., & Zacks, S. (2015). Exact calculation of the OC and ASN of a truncated SPRT for the mean of a exponential distribution. *Methodology and Computing in Applied Probability, 17*, 915–927.

Dempster, A. P., Laird, N. M., & Rubin, D. B. (1977). Maximum likelihood from incomplete data via EM algorithm. *Journal of the Royal Statistical Society, B, 39*, 1–38.

Di Crescenzo, A., Iuliano, A., Martinucci, B., & Zacks, S. (2013). Generalized telegraph process with random jumps. *Journal of Applied Probability, 50*, 450–463.

Di Crescenzo, A., Martinucci, B., & Zacks, S. (2015). Compound Poisson with Poisson subordinator. *Journal of Applied Probability, 52*, 360–374.

Di Crescenzo, A., Martinucci, B., & Zacks, S. (2018). Telegrapher process with elastic boundary at the origin. *Methodology and Computing in Applied Probability, 20*, 333–352.

Donoho, D. (2017). 50 years of data science. *Journal of Computational and Graphical Statistics, 26*, 745–766.

Durham, S. D., & Flournoy, N. (1995). Up-and down designs I: Stationary treatment distributions. *Adaptive Designs IMS Lecture Notes - Monograph Series, 25*, 139–157.

Efron, B., & Hastie, T. (2016). *Computer age statistical inference, algorithms, evidence, and data science*. New York: Cambridge University Press.

Ehrenfeld, S., & Zacks, S. (1961). Randomization and factorial experiments. *Annals of Mathematical Statistics, 32*, 270–297.

Ehrenfeld, S., & Zacks, S. (1963). Optimal strategies in factorial experiments. *Annals of Mathematical Statistics, 34*, 780–791.

Ehrenfeld, S., & Zacks, S. (1967). Testing hypotheses in randomized factorial experiments. *Annals of Mathematical Statistics, 38*, 1494–1507.

Eichhorn, B. H. (1974). Sequential search of an optimal dosage for cases of linear dosage-toxicity regression. *Communications in Statistics, 3*, 263–271.

Eichhorn, B. H., & Zacks, S. (1973). Sequential search of an optimal dosage I. *Journal of the American Statistical Association, 68*, 594–598.

Eichhorn, B. H., & Zacks, S. (1981). Bayes sequential search of an optimal dosage: Linear regression with both parameters unknown. *Communications in Statistics - Theory and Methods, 10*, 931–953.

Finney, D. J. (1978). *Statistical method in biological assay. Mathematics in medicine series* (3rd ed.). London: Hodder Arnold.

Fisher, R. A. (1925). *Statistical methods for research workers*. Edinburgh: Oliver and Boyed.

Freeman, P. R. (1972). Sequential estimation of the size of a finite population. *Biometrika, 59*, 9–17.

Frenkel, J., & Zacks, S. (1957). Wind-produced energy and its relation to wind regime. *Bulletin of the Research Council of Israel, 6A*, 189–194.

George, E. (1991). Shrinkage domination in a multivariate common mean problem. *The Annals of Statistics, 19*, 952–960.

Ghezzi, D., & Zacks, S. (2005). Inference on the common variance of correlated normal random variables. *Communication in Statistics-Theory and Methods, 34*, 1517–1531.

Ghosh, M., Mukhopadhyay, N., & Sen, P. K. (1997). *Sequential estimation*. New York: Wiley.

Graybill, F. A., & Deal, R. B. (1959). Combining unbiased estimators. *Biometrics, 15*, 543–550.

Graybill, F. A., & Weeks, D. L. (1959). Combining inter-block and intra-block in incomplete blocks. *Annals of Mathematical Statistics, 30*, 799–805.

Gut, A. (1988). *Stopped random walks: Limit theorems and applications*. New York: Springer.

Hald, A. (1952). *Statistical theory with engineering applications*. New York: Wiley.

Hanner, D. M., & Zacks, S. (2013). On two-stage sampling for fixed-width interval estimation of the common variance of equi-correlated normal distributions. *Sequential Analysis, 32*, 1–13.

Hedayat, A. S., & Sinha, B. K. (1991). *Design and inference in finite populations sampling*. New York: Wiley.

James, W., & Stein, C. (1960). Estimation with quadratic loss. In *Proceedings of Fourth Berkeley Symposium on Mathematical Statistics and Probability* (vol. 1). Berkely: Statistics Laboratories of the University of California.

Johnson, N. L., & Kotz, S. (1969). *Distributions in statistics: Discrete distributions*. Boston: Houghton and Mifflin.

Kaplan, D. (2018). Teaching stats for data science. *The American Statistician, 7*, 89–96.

Keller, T., & Olkin, I. (2004). Combining correlated unbiased estimators of the mean of a Normal distribution. *A Festschrift for Herman Rubin, IMS Lecture Notes-Monographs, Series, 45*, 218–227.

Kenett, R. S., & Redman, T. C. (2019). *The real work of data science: Turning data into information, better decisions, and strong organizations*. New York: Wiley.

Kenett, R. S., & Zacks, S. (1998). *Modern industrial statistics: Design and control of quality and reliability*. Belmont: Duxbury.

Kenett, R. S., & Zacks, S. (2014). *Modern industrial statistics with R, minitab and JMP* (2nd ed.). Chichester: Wiley.

Kubokawa, T. (1987a). Estimation of the common mean of normal distributions with applications to regression and design of experiments. Ph.D. Dissertation, University of Tsukuba.

Kubokawa, T. (1987b). Admissible minimax estimators of a common mean of two normal populations. *Annals of Statistics, 15*, 1245–1256.

Lehmann, E. L. (1959). *Testing statistical hypotheses*. New York: Wiley.

Leite, J. G., & Pereira, C. A. (1987). An urn model for the capture-recapture sequential sampling process. *Sequential Analysis, 6*, 179–186.

Lorden, G. (1971). Procedures for reacting to a change in distribution. *The Annals of Mathematical Statistics, 42*, 1897–190.

Mood, A. M. (1946). On Hotelling's weighing problems. *Annals of Mathematical Statistics, 17*, 432–446.

Morris, C. (1983). Parametric empirical Bayes inference: Theory and applications. *Journal of the American Statistical Association, 78*, 47–54.

Mukhopadhyay, N., & Cicconetti, G. (2004). Estimating reliabilities following purely sequential sampling from exponential populations. In *Advances on ranking and selection, multiple comparisons and reliability* (pp. 303–332). Boston: Birkhauser.

Page, E. S. (1954). Continuous inspection schemes. *Biometrika, 41*, 100–114.

Page, E. S. (1955). A test for a change in a parameter occurring at an unknown point. *Biometrika, 42*, 523–527.

Page, E. E. (1957). On problems in which a change of parameter occurs at an unknown point. *Biometrika, 44*, 248–252.

Perry, D., Stadje, W., & Zacks, S. (1999). First- exit times for increasing compound processes. *Communications in Statistics-Stochastic Models, 15*, 977–992.

Perry, D., Stadje, W., & Zacks, S. (2002). Hitting and ruin probabilities for compound Poisson processes and the cycle maximum of the M/G/1 queue. *Stochastic Models, 18*, 553–564.

Peskir, G., & Shiryaev, A. N. (2002). Solving the Poisson disorder problem. In K. Sandman & P. J. Schonbucher (Eds.). *Advances in finance and stochastics* (pp. 295–312). New York: Springer.

Plunchenko, A. S., & Tartakovsky, A. G. (2010). On optimality of Shiryaev-Roberts procedure for detecting a change in distributions. *The Annals of Statistics, 39*, 3445–3457.

Pollak, M. (1985). Optimal detection of a change in distribution. *The Annals of Statistics, 15*, 749–779.

Raj, D. (1972). *The design of sampling surveys*. New York: McGraw Hill.

Rogatko, A., Babb, J. S., & Zacks, S. (1998). Cancer phase I clinical trials: Efficient dose escalation with over dose control. *Statistics in Medicine, 17*, 1103–1120.

Seber, G. A. F. (1985). *The estimation of animal abundance* (2nd ed.). London: Griffin.

Shewhart, W. A. (1931). *Economic control of quality of manufactured product.* Princeton: Van Nostrand.

Shiryaev, A. N. (1963). On optimum methods in quickest detection problem. *Theory of Probability and its Applications, 7*, 22–46.

Snedecor, G. W., & Cochran, W. G. (1980). *Statistical methods* (7th ed.). Ames: Iowa State University Press.

Stadje, W., & Zacks, S. (2003). Upper first-exit times of compound Poisson processes revisited. *Probability in the Engineering and Informational Sciences, 17*, 459–465.

Stadje, W., & Zacks, S. (2004). Telegraph process with random velocities. *Journal of Applied Probability, 41*, 665–678.

Starr, N., & Woodroofe, M. (1972). Further remarks on sequential estimation: The exponential case. *Annals of Mathematical Statistics, 43*, 1147–1154.

Woodroofe, M. (1977). Second order approximation for sequential point and interval estimation. *The Annals of Statistics, 48*, 984–995.

Woodroofe, M. (1982). *Non-linear renewal theory in sequential analysis.* Philadelphia: SIAM.

Yadin, M., & Zacks, S. (1982). Random coverage of a circle with application to a shadowing problem. *Journal of Applied Probability, 19*, 562–577.

Yadin, M., & Zacks, S. (1985). The visibility of stationary and moving targets in the plane subject to a Poisson field of shadowing elements. *Journal of Applied Probability, 22*, 76–786.

Yadin, M., & Zacks, S. (1988). Visibility probabilities on line segments in three dimensional spaces subject to a random Poisson field of observing spheres. *Naval Research Logistics Quarterly, 35*, 555–569.

Yadin, M., & Zacks, S. (1994). The survival probability function of a target moving along a straight line in a random field of obscuring elements. *Naval Research Logistics Quarterly, 41*, 689–706.

Zacks, S. (1963a). On a complete class of linear unbiased estimators for randomized factorial experiments. *The Annals of Mathematical Statistics, 34*, 769–779.

Zacks, S. (1963b). Optimal strategies in factorial experiments. *The Annals of Mathematical Statistics, 34*, 780–791.

Zacks, S. (1964). Generalized least squares estimators for randomized fractional replication designs. *The Annals of Mathematical Statistics, 35*, 696–704.

Zacks, S. (1966a). Randomized fractional weighing designs. *The Annals of Mathematical Statistics, 37*, 1382–1395.

Zacks, S. (1966b). Unbiased estimation of the common mean of two normal distributions based on small samples. *Journal of the American Statistical Association, 61*, 467–476.

Zacks, S. (1967). Bayes sequential strategies for crossing a field containing absorption points. *Naval Research Logistics Quarterly, 14*, 329–343.

Zacks, S. (1968). Bayes sequential design of fractional factorial experiments for the estimation of a subgroup of pre-assigned parameters. *The Annals of Mathematical Statistics, 39*, 973–982.

Zacks, S. (1969a). Bayes sequential design of stock levels. *Naval Research Logistics Quarterly, 16*, 143–155.

Zacks, S. (1969b). Bayes sequential design of fixed samples from finite populations. *Journal of the American Statistical Association, 64*, 1342–1349.

Zacks, S. (1970a). Bayesian design of single and double stratified sampling for estimating proportion finite populations. *Technometrics, 12*, 119–130.

Zacks, S. (1970b). Bayes and fiducial equivariant estimators of the common mean to two normal distributions. *The Annals of Mathematical Statistics, 41*, 59–67.

Zacks, S. (1970c). A two-echelon multi-station inventory model for navy applications. *Naval Research Logistics Quarterly, 17*, 79–85.

Zacks, S. (1974). On the optimality of the Bayes prediction policy in two-echelon multi-station inventory systems. *Naval Research Logistics Quarterly, 21*, 569–574.

Zacks, S. (1979). Survival distributions in crossing fields containing clusters of absorption points with possible detection and uncertain activation or kill. *Naval Research Logistics Quarterly, 26*, 423–435.

Zacks, S. (1981a). Bayes and equivariant estimators of the variance of a finite population, I: Simple random sampling. *Communications in Statistics - Theory and Methods, A10*, 407–426.

Zacks, S. (1981b). Bayes equivariant estimators of the variance of a finite population for exponential priors. *Communications in Statistics - Theory and Methods, A10*, 427–437.

Zacks, S. (1983). Survey of classical and Bayesian approaches to the change point problem: Fixed sample and sequential procedures of testing and estimation. In D. Siegmund, J. Rustagi, & G. Gaseb Rizvi (Eds.). *Recent advances in statistics* (pp. 245–269). New York: Academic.

Zacks, S. (1984). Estimating the shift to wear-out of systems having Exponential-Weibull life distributions. *Operations Research, 32*, 741–749.

Zacks S. (1985). Distribution of stopping variables in sequential procedures for the detection of shifts in the distributions of discrete random variables. *Communications in Statistics, B9*, 1–8.

Zacks, S. (1988), A simulation study of the efficiency of Empirical Bayes' estimators of multiple correlated probability vectors. *Naval Research Logistics Quarterly, 35*, 237–246.

Zacks, S. (1991). Distributions of stopping times for Poisson processes with linear boundaries. *Communication in Statistics. Stochastic Models, 7*, 233–242.

Zacks, S. (1992). *Introduction to reliability analysis: Probability models and statistical methods.* New York: Springer.

Zacks, S. (1994). *Stochastic visibility in random fields. Lecture notes in statistics* (vol. 95). New York: Springer.

Zacks, S. (2002). In the footsteps of Basu: The predictive modeling approach to sampling from finite population. *Sankhya, Series A, 64*, 532–544.

Zacks, S. (2004). Distribution of failure times associated with non-homogeneous compound Poisson damage processes. *IMS Lecture Notes, 45*, 396–407.

Zacks, S. (2009). *Stage-wise adaptive designs.* New York: Wiley.

Zacks, S. (2010). The availability and hazard of a system. *Methodology and Computing in Applied Probability, 12*, 555–565.

Zacks, S. (2012). Distribution of the total time in a mode of an alternating renewal processes with applications. *Sequential Analysis, 31*, 397–408.

Zacks, S. (2014). *Problems and examples in mathematical statistics.* New York: Wiley.

Zacks, S. (2017). *Sample path analysis and boundary crossing times. Lecture notes in mathematics* (vol. 2203). Berlin: Springer.

Zacks, S., Babb, J. S., & Rogatko, A. (1998). Bayes sequential search of an optimal dosage. *Statistics and Probability Letters, 38*, 215–220.

Zacks, S., & Barzily, Z. (1981). Bayes procedures for detecting a shift in the probability of success in a series of Bernoulli trials. *Journal of Statistics Planning & Inference, 5*, 107–119.

Zacks, S., & Fennel, J. (1972). Bayes adaptive control of two-echelon inventory systems, I: Development for a special case of one-station lower echelon and Monte Carlo evaluation. *Naval Research Logistics Quarterly, 19*, 15–28.

Zacks, S., & Fennel, J. (1973). Distribution of adjusted stock levels under statistical adaptive control procedures for inventory systems. *Journal of the American Statistical Association, 68*, 88–91.

Zacks, S., & Fennel, J. (1974). Bayes adaptive control of two-echelon inventory systems, II: The multi-station case. *Naval Research Logistics Quarterly, 21*, 575–593.

Zacks, S., & Goldfarb, D. (1966). Survival probabilities in crossing a field containing absorption points. *Naval Research Logistics Quarterly, 13*, 35–48.

Zacks, S., & Mukhopadhyay, N. (2006). Exact risk of sequential point estimators of the exponential parameters. *Sequential Analysis, 25*, 203–226.

Zacks, S., & Ramig, P. A. (1987). Confidence intervals for the common variance of equi-correlated normal random variables. In A. E. Gelfand (Ed.) *Contributions for the theory and applications of statistics* (pp. 511–544). A Volume in Honor of Herbert Solomon. New York: Academic.

Zacks, S., & Rodriguez, J. (1986). A note on the missing value principle and the EM-algorithm for estimation and prediction in sampling from finite populations with a multinormal superpopulation model. *Statistics & Probability Letters, 4*, 35–37.

Zacks, S., & Solomon, H. (1981). Bayes equivariant estimators of the variance of a finite population: Part I, simple random sampling. *Communication in Statistics, Theory and Methods, 10*, 407–426.

Zacks, S., & Yadin, M. (1984). The distribution of the random lighted portion of a curve in a plane shadowed by a Poisson random field of obstacles. In E. J. Wegman & J. G. Smith (Eds.), *Statistical signal processing*. New York: Marcel Dekker.

Zacks, S., Marlow, W. H., & Breir, S. S. (1985). Statistical analysis of very high-dimensional data sets of hierarchically structured binary random variables with missing data and application to Marine Corps readiness evaluations. *Naval Research Logistics Quarterly, 32*, 467–490.

Zacks, S., Pereira, C.A., & Leite, J. G. (1990). Bayes sequential estimation of the size of a finite population. *Journal of Statistical Planning and Inference, 25*, 363–380.

Zacks, S., Perry, D., Bshuty, D., & Bar-Lev, S. (1999). Distributions of stopping times for compound Poisson processes with positive jumps and linear boundaries. *Communications in Statistics. Stochastic Models, 15*, 89–101.

Author Index

© Springer Nature Switzerland AG 2020
S. Zacks, *The Career of a Research Statistician*, Statistics for Industry,
Technology, and Engineering, https://doi.org/10.1007/978-3-030-39434-9

Subject Index

Printed in the United States
by Baker & Taylor Publisher Services